EAI/Springer Innovations in Communication and Computing

Editor's Note

The impact of information technologies is creating a new world yet not fully understood. The extent and speed of economic, life style and social changes already perceived in everyday life is hard to estimate without understanding the technological driving forces behind it. This series presents contributed volumes featuring the latest research and development in the various information engineering technologies that play a key role in this process.

The range of topics, focusing primarily on communications and computing engineering include, but are not limited to, wireless networks; mobile communication, design and learning; gaming; interaction; e-health and pervasive healthcare; energy management; smart grids; internet of things; cognitive radio networks; computation; cloud computing; ubiquitous connectivity, and in mode general smart living, smart cities, Internet of Things and more. The series publishes a combination of expanded papers selected from hosted and sponsored European Alliance for Innovation (EAI) conferences that present cutting edge, global research as well as provide new perspectives on traditional related engineering fields. This content, complemented with open calls for contribution of book titles and individual chapters, together maintain Springer's and EAI's high standards of academic excellence. The audience for the books consists of researchers, industry professionals, advanced level students as well as practitioners in related fields of activity include information and communication specialists, security experts, economists, urban planners, doctors, and in general representatives in all those walks of life affected ad contributing to the information revolution.

Indexing: This series is indexed in Scopus, Ei Compendex, and zbMATH.

About EAI

EAI is a grassroots member organization initiated through cooperation between businesses, public, private and government organizations to address the global challenges of Europe's future competitiveness and link the European Research community with its counterparts around the globe. EAI reaches out to hundreds of thousands of individual subscribers on all continents and collaborates with an institutional member base including Fortune 500 companies, government organizations, and educational institutions, provide a free research and innovation platform.

Through its open free membership model EAI promotes a new research and innovation culture based on collaboration, connectivity and recognition of excellence by community.

More information about this series at http://www.springer.com/series/15427

M. Kathiresh • R. Neelaveni
Editors

Automotive Embedded Systems

Key Technologies, Innovations, and Applications

Springer

Editors
M. Kathiresh
Department of Electrical and Electronics Engineering
PSG College of Technology
Coimbatore, Tamil Nadu, India

R. Neelaveni
Department of Electrical and Electronics Engineering
PSG College of Technology
Coimbatore, Tamil Nadu, India

ISSN 2522-8595 ISSN 2522-8609 (electronic)
EAI/Springer Innovations in Communication and Computing
ISBN 978-3-030-59896-9 ISBN 978-3-030-59897-6 (eBook)
https://doi.org/10.1007/978-3-030-59897-6

This Springer imprint is published by the registered company Springer Nature Switzerland AG
The registered company address is: Gewerbestrasse 11, 6330 Cham, Switzerland

Preface

Over the last two decades, various functions of a vehicle are performed by using electrical and electromechanical systems, which were performed by mechanical linkages in conventional automotive systems. The concept of using electronic control systems as replacement for mechanical control systems in automobiles is called as drive-by-wire or x-by-wire. The main functions such as acceleration, braking, and steering are controlled by the use of mechanical, pneumatic, and hydraulic components in conventional vehicles where are more prone to wear and tear. Thus, the efficiency and performance of such vehicles deteriorates over a period of time. In contrast, the drive-by-wire technology uses sensors, electrical motors, and electromechanical actuators to perform vehicular functions. Moreover, the subsystems in the drive-by-wire technology have a dedicated Electronic Control Unit (ECU) to monitor and control vehicle parameters with the help of appropriate sensors and actuators. These subsystems are called as Automotive Embedded Systems. Embedded systems in automobiles are basically classified into five domains such as Power Train, Body Electronics, Chassis, Human–Machine Interface, and Telematics.

The main objective of Industry 4.0, the Fourth Industrial Revolution is to make everything smart and connected with each other. The tremendous growth in automotive electronics and wireless communication technology has paved a way for new technology called Connected Cars through which many innovative features have been added in a typical car to enhance the comfort of the stake holders.

This book starts with automotive safety systems which is one of the major functional domains. The book discusses the importance of software in automotive systems followed by an insight into Automotive Software Standards, MISRA Coding Standards, and Model-based Software Development Approach. The book further discusses vehicle diagnostics and over-the-air software update processes. The book also illustrates the role of sensors and artificial intelligence in automotive systems. Various innovative applications involving the concept of Internet of Things are also presented in this book. This book is intended for academicians, researchers, and industrialists.

Coimbatore, Tamil Nadu, India — M. Kathiresh
Coimbatore, Tamil Nadu, India — R. Neelaveni

Acknowledgment

We would like to thank Mr. Imrich Chlamtac, the Series Editor and Ms. Eliska, the Managing Editor at European Alliance for Innovation, for giving us this opportunity to edit a book in the Series of Innovations in Communication and Computing, Springer. We express our sincere thanks to the authors for their contribution and whole-hearted cooperation in the making of this book. We also thank the reviewers for their constructive criticism and comments which enriched this work. We are grateful to PSG College of Technology, Coimbatore, India, for the constant support and encouragement. Finally, we would like to acknowledge with gratitude, the support, and love of our family members and colleagues, without whom this book would not have been possible.

Contents

About the Editors

M. Kathiresh is a Faculty in the Department of Electrical and Electronics Engineering, PSG College of Technology, Coimbatore, India. He has completed his Under Graduation in the stream of Electronics and Communication Engineering, Post-Graduation in Embedded and Real-Time Systems and Doctorate in the area of Automotive Embedded Systems at Anna University, Chennai. He has done several innovative projects which are now published in reputed journals, international conferences, books, and some are patented. He has also carried out many consultancy projects for various leading industries to help the industries to adapt for Industry 4.0. He is currently working in a research project funded by the Department of Science and Technology, India. His research areas include Automotive Embedded Systems, Internet of Things, and Real-Time Embedded Systems. He is a life member of Institute of Engineers (India) and Indian Society of Technical Education.

R. Neelaveni is presently working as Professor in the Department of Electrical and Electronics Engineering, PSG College of Technology, Coimbatore, India. She completed her Bachelor Degree in Electronics and Communication Engineering (Madras University) and Masters in Applied Electronics (Bharathiar University) from PSG College of Technology, Coimbatore. She obtained her Ph.D. in Biomedical Instrumentation from Bharathiar University, Coimbatore. She has around 31 years of teaching experience. Her research interests include Bio-signal processing, Bio-image processing, Computer Networks, Soft computing, and Embedded Systems. She was awarded the Best National Service Scheme Programme Officer of Anna University, Chennai in the year 2004–2005. She has served as member of Board of Studies in autonomous colleges and panel member for faculty selection in various colleges. She has published many papers in various reputed International Journals and Conferences. She also holds patents and has contributed chapters in books. Currently, she is working in a research project funded by the Department of Science and Technology, India. She is a life member of Indian Society of Technical Education.

Automotive Safety Systems

S. Hamsini and M. Kathiresh

Introduction

Road traffic crashes lead to an estimated death of nearly 1.25 million people in a year and millions more succumb to injuries [1]. In order to reduce such gruesome road traffic fatalities and injuries, various measures grounded on evidence have been formulated by the World Health Organization. A critical role is played by safe vehicles to avert crashes and minimize the probability of serious injury. Safety standards for automobiles are set by the United Nations World Forum for Harmonization of Vehicle Regulations along with the legal framework for which voluntary application by member states is possible. Vehicles that abide to these safety standards are less prone to accidents, and on the occurrence of road traffic crashes, the injuries are far from life-threatening [2].

Though there are various contributing factors to accidents, the major single reason points towards the driver as human errors contribute to a major percentile of accidents. Modern vehicles include various safety features that assist the driver in various possible ways. The modern vehicle design is focused on better safety features for the driver and the passengers.

The systems that get activated in retribution to an abnormal event such as a safety problem are termed as active safety systems. These systems are triggered in anomalous circumstances by human operators or spontaneously by an intelligent computer system. The systems that act in a favorable manner in response to dangerous events by relying on the laws of nature are named as passive safety systems. This safety

S. Hamsini (✉)
Robert Bosch Engineering and Business Solutions Private Limited, Coimbatore, India

M. Kathiresh
PSG College of Technology, Coimbatore, India

M. Kathiresh, R. Neelaveni (eds.), *Automotive Embedded Systems*,
EAI/Springer Innovations in Communication and Computing,
https://doi.org/10.1007/978-3-030-59897-6_1

system may be a physical feature or an added internal system or design modifications that enhance the driving of the vehicle.

Vehicle Safety Systems

When the design of a vehicle is considered, one amongst the various important aspects is its safety both in the vehicle as well as in its features. Automobile safety is the scientific domain related to the study, design, construction, and regulation of technology mainly focused on minimizing the occurrence and consequences of road traffic accidents. Vehicle occupant safety can be generally categorized into two areas: firstly, active safety, which is basically designed for crash avoidance, and secondly, passive safety also known as crash worthiness where in the worst case that a crash happens, protection of occupant is ensured [3]. The various active safety and passive safety features are depicted in Fig. 1.

The assistance provided by technology for the prevention of a crash is termed as active safety, whereas the various physical components of an automobile that collectively help to protect the inhabitants when a crash occurs are referred as passive safety. Figure 2 illustrates the timeline pertaining to the active safety and passive safety.

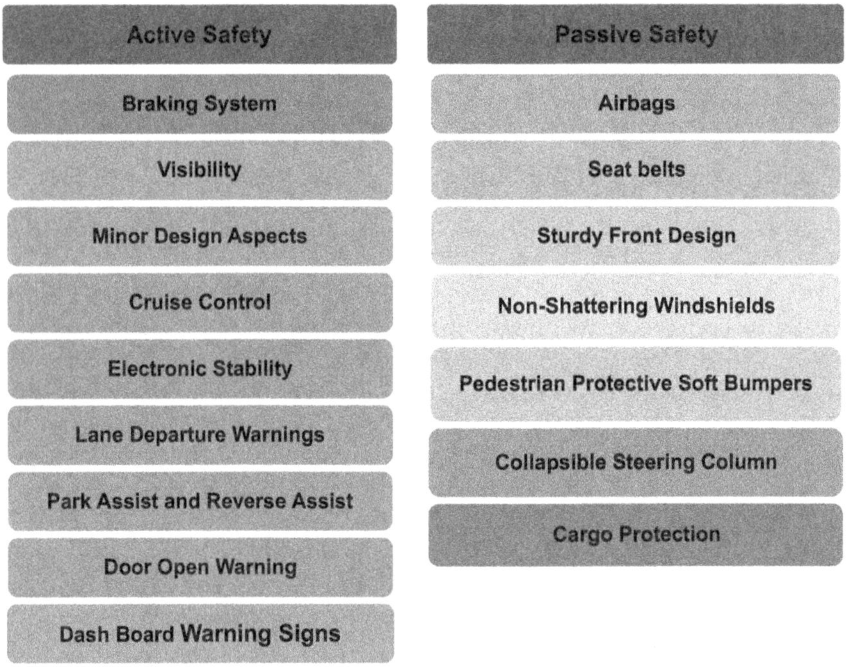

Fig. 1 Active and passive safety features

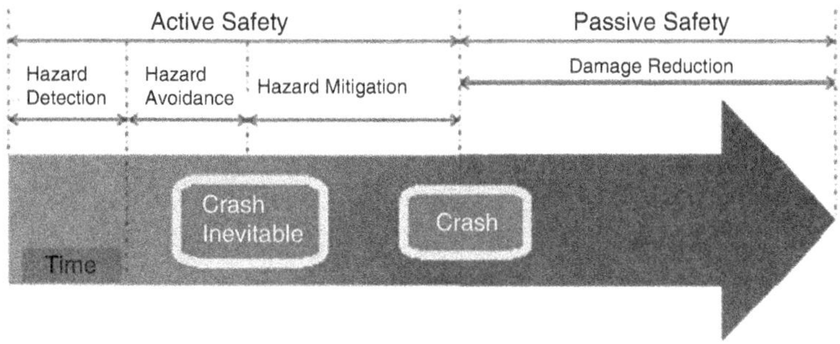

Fig. 2 Vehicle crash timeline of safety

Active Safety Systems

The features of a vehicle that help to mitigate or prevent road crashes are basically active safety features. These features mainly focus on prevention of the crash or reduction of severity of a crash that is unavoidable in nature. One or more aspects of an automobile are monitored constantly by the active safety features for potential hazards. In the instance where a problem occurs, the situation is rectified autonomously by active safety features. Protection that is offered is increased by active safety features. The devices and systems used for active safety are generally automated. Few basic active safety features are discussed below.

Braking System

A brake is basically a mechanical device that operates by inhibiting motion by means of absorbing energy from a moving object. It is generally intended for decelerating or altogether stopping an automobile in motion, its wheels and axle, and this is frequently accomplished through friction. The rapid advancements in vehicles and modern road infrastructure facilitate faster driving habits. In this scenario, when hard brake is applied to a moving vehicle, the wheels may get locked up and end up sending the vehicle to spin out of control. In order to avoid those mishaps, an anti-skid system involving the brake was developed which is now improved as Anti-Lock Braking System (ABS).

Movement of vehicle is the result of friction between the vehicle's tyres and the road. A torque is applied in the opposite direction to that of the friction, thus ultimately leading to the wheel halting the rotations. While braking, ABS allows the vehicle's wheels to maintain tractive contact based on driver inputs with the road surface. Avoidance of uncontrolled skidding and prevention of locking up of wheels are attained.

The principles of threshold braking and cadence braking that were in practice by experienced drivers formerly are employed in the ABS as an automated system. Due to the automation, ABS acts in a considerably faster rate and in a very effective manner than manual operation. Enhanced vehicle control and reduced stopping distances is provided by the ABS on dry and certain slippery surfaces, but the performance is comparatively diminished, and the braking distance may increase on loose gravel roads or surfaces concealed with snow. Initially ABS was introduced for production vehicles; hence the sophistication and effectiveness of these systems have increased. Not only the prevention of wheel lock under braking is attained but also alteration of the front-to-rear brake bias is accomplished. Illumination of a warning light located on the vehicle instrument panel indicates a fault within the ABS. Until the fault is resolved, the ABS will be incapacitated. The modern ABS involves a dedicated microcontroller controlling all the four wheels by means of a control system consisting of hub-mounted sensors to apply individual brake pressure. Due to rapid escalation of popularity of the electronic stability systems due to the massive reduction in cost of vehicular electronics in the recent years, the ABS is present as a standard entity in most of the vehicles nowadays. A study has proved that automobiles with ABS are less likely to go through fatal accidents by a percentage of 37%.

Brake lights are another feature of braking that is relatively less technical and more of a handy feature while driving. Brake lights are simple indicating lights usually red to enhance visibility even in daytime. These lights glow brighter, indicating the vehicle following that the brakes are applied and the vehicle is going to decelerate or stop. It is more of a traffic feature that facilitates fellow drivers to adjust to the traffic when someone applies brakes.

Parking brakes also known as hand brakes are another feature that is used to secure the vehicle motionless when parked. The general idea is to restrict the uncontrolled movement of the vehicle due to slope or external impacts when parked, but historically it also acted as a secondary emergency brake in situations where the main brake fails. But with improved braking systems like ABS, the parking brakes are needed only when the vehicle is parked.

As seen with the ABS and cruise control, the brake assist system helps a driver in many ways from applying emergency brake pressure to preventing collision and skidding. The modern brake assist is integrated with an anti-collision system where brakes will apply automatically on detecting a possibility of collision. It also gives warnings about the health of brake system.

Visibility

Visibility to the driver is a major factor while plying on roads. Features including rearview mirrors, wipers, and clear windshields enhance the visibility for the driver to see.

Seating position of the driver is a key parameter as it decides the visibility of the road and surroundings to the driver. One should choose minimum or no blind spot position to sit while driving. Many features like adjustable seat position, height, and adjustable steering columns help in achieving a comfortable seating position for driving.

Though the feature of automatic climate control is more of a comfort feature to control temperature and humidity inside the vehicle, it also plays an important role in preventing formation of fog in windows and windshields. Thus, it improves visibility during winters and rainy season.

Rearview mirrors give access for the driver to see the sides and rear side of the vehicle. It helps the driver to decide before changing lanes or attempting a turn and helps to monitor the entire vehicle. Advanced rearview mirrors also have features like electronic adjustments to counter parallax errors and some vehicles have camera sensors that capture the video and display it in the screen inside the vehicle.

Headlamps facilitate the drivers to drive the vehicle during nights. However, advanced features in headlamps such as adaptive focus and brightness facilitate the driver with a better and comfortable visibility [4].

Wipers are a simple link mechanism that helps in wiping the water in the windshield while raining. During rains, it is difficult for the driver to get proper visibility due to water creating refraction with the windshield glass. Advanced features in wipers include automatic rain sensing where the wiper gets activated on sensing raindrops on the windshield.

There is another aspect of visibility where the vehicle needs to have some added features to have better visibility to be seen by others.

Recent standards for vehicles after BS IV and Euro IV made it mandatory for the vehicles to have running headlamps even during the day. The idea is mainly to make the vehicle more visible even from a distance. With advancements in headlamps after the introduction of LEDs and their minimal cost for installation and maintenance, it is quite a handy feature of the vehicle.

Parking lights are used when the vehicle is pulled over in highways or some work is happening around the vehicle that is parked. It should not be used when the vehicle is moving.

The color of a car has been a debatable study for the past decades as to how it plays a role in road accidents. Some of the latest studies essentially suggest that cars with brighter colors such as white are relatively less prone to accidents than the ones with black or gray. Apart from the study being under scrutiny to be proven factual, it is a general idea that a better visible color enhances the visibility of the vehicle.

Another significant instance when it necessitates to get the attention of fellow drivers on the road is especially when taking turns or going reverse, and the accessories that aid for this are turning indicators and reverse signal indicators. These indicators give the signals by glowing intermittently, sometimes combined with a buzzer or tune to alert the presence to fellow drivers.

Horns and tunes are used to grab the attention of fellow drivers to indicate the action to be performed by the vehicle.

Minor Design Aspects

Bumpers fitted in front and the rear of the vehicle absorbs impacts during minor collisions, preserving internal major components and thereby reducing repair cost. However modern safety standards suggest using soft bumpers to enhance pedestrian safety.

The intended design function of a spoiler which is an aerodynamic device is to "spoil" hostile air movement across a moving vehicle, generally termed as drag or turbulence. Spoilers employed in the front of an automobile are termed as air dams. Spoilers are frequently attached to sports cars of high performance and to vehicles intended for racing, though nowadays it is a common accessory in passenger cars too. Few spoilers that are employed in the cars are predominantly for styling purposes and aid very less towards aerodynamics. The wing in a vehicle refers to a device intended to generate downforce when air passage occurs around it and to not disrupt the existing patterns of air flow. Thus, instead of decreasing the drag, automotive wings in reality increase drag.

Cruise Control

Cruise control also known as speed control automatically maintains the speed of the vehicle by taking over the throttle of the vehicle and maintains steady speed. This feature is especially helpful on highways and expressways where a steady speed needs to be maintained over a long period. Figure 3 illustrates the operation of cruise control system present in vehicles.

More advanced adaptive cruise control mechanisms integrate features like radar-based dynamic cruise control and automatic braking system on detecting potential forward collision, and this is essentially a principal step towards autonomous vehicles where a vehicle varies its speed through intelligent radar-based sensor system

Fig. 3 Cruise control

and also stops on detecting collision probability with near-zero human interventions [5].

Speed governors are actually an optional feature which restricts the vehicle from going beyond certain prescribed speed limits.

Autopilot is an advanced active safety feature envisioned for advanced driver assistance to enhance the safety and convenience of the driver behind the wheel. A few autopilot systems include a suite of driver assistance features like traffic aware cruise control, auto lane change, summoning where the car navigates in a parking lot to an intended location while maneuvering around objects that hinder its navigation otherwise and full self-driving capability among various other features.

Stability

Physical stability of the vehicle is more of a design aspect of the vehicle that provides mechanical stability to the vehicle while driving. Proper wheel balancing, sturdy and rigid chassis, optimum tire pressure, distribution of body weight, steering system, and suspensions majorly contribute to the physical stability of the vehicle. Stringent standards and guidelines are provided for the manufacturers to ensure these design aspects [6].

Electronic stability control (ESC) is a computerized technology which is intended to detect and reduce loss of traction that results in skidding and thus overall improves the stability of a vehicle, and it is also known as dynamic stability control (DSC) or electronic stability program (ESP). Brakes are applied automatically when loss of steering control is detected by ESC, and it steers the vehicle where it is supposed to go according to the driver's intention. Braking is applied automatically to the wheels separately, namely, the outer wheel in the front as a counter for steer and the inner wheel in the rear to handle understeer. Engine power is also reduced by a few ESC systems till control is attained. The cornering performance of the vehicle is not improved by ESC, but the loss of control is minimized with its help. ESC is a computerized system that involves various sensors such as wheel speed sensors, steering angle sensor, accelerometer, etc.

A vehicle can be steered when it veers or slides out of the intended path due to natural hindrances such as rain or snow. The path of a vehicle in both the instances where ESC is present and not when loss of traction occurs is depicted in the Fig. 4.

Traction control system is an active vehicle safety feature which is actually a secondary feature of electronic stability control and kicks in when the vehicle's speed increases and prevents loss of traction of the wheels driven on roads. This active safety feature is activated while driving when the throttle input and the engine torque do not match.

Other key stability warnings are tyre deflation warnings which prompts the driver that the tyres are either deflated or not in the optimum pressure for the drive and air suspension warning system which essentially warns about faulty functioning of suspension due to various reasons.

Fig. 4 Electronic stability control

Fig. 5 Lane departure warnings

Lane Departure Warnings

A lane departure warning system has the intended purpose of monitoring the lane markings and detecting the drifting or veering of vehicles out of its own lane. These systems then warn the drivers through an audio alert or a visual illustration regarding the unintentional lane shifts and to navigate the automobile back into its intended lane. Figure 5 represents the lane departure warnings that arise when the driver changes lanes along with the outline of the current lane where the vehicle is navigating.

Park Assist and Reverse Assist

The most complicated aspect of driving involves parking of the vehicle properly in a narrow space. Only skillful drivers can perform this parking without much difficulty. Therefore, an assist to facilitate all the drivers to park the vehicle is introduced in modern cars. The intelligent park assist can steer itself into a parking space with no or minimum input from the driver eliminating minor accidents. Another notable feature is where the reverse assist essentially helps the driver to park when the vehicle is put on reverse. It is integrated with a reverse camera and sensors, and the view is displayed to the driver through a screen that enables him to perform complex reverse maneuvers easily. There are various sensors such as ultrasonic sensors to detect the proximity of various objects within the vicinity of the vehicle in the rear and warn about it.

Door Open Warning

Door open warning systems encompass a sensor that detects any approaching vehicles in the rear when an individual is opening the car door and prevents the opening of the car door when vehicles are detected within the vicinity and also warns the approaching vehicle by means of cautioning lights and buzzers. The door open warning system is depicted in Fig. 6

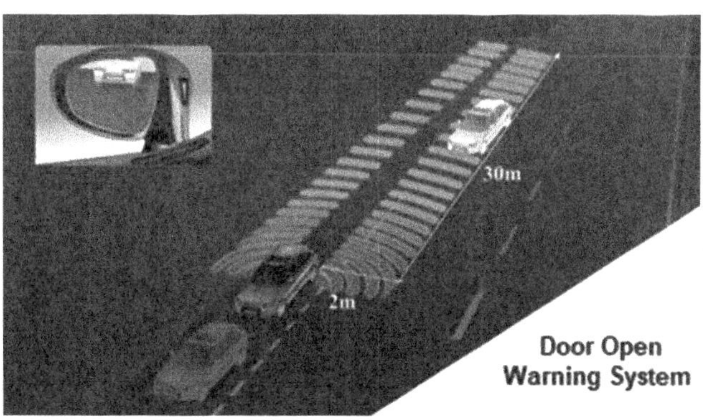

Fig. 6 Door open warning

Dashboard Warning Signs for Various Aspects of ECU and Component Failure

There are a lot of other warning signs in the dashboard that facilitate the driver to have a tab over various aspects of the vehicle. Key features include door not closed warning sign, low fuel indicators, seat belt warnings, overheating of component or engine warnings, headlamp status, indicator, and much more. Some the warning features in the dashboard in a modern vehicle are illustrated as in Fig. 7

Passive Safety Systems

Apart from various active safety features, there are many passive safety features of the car which is also known as crashworthiness of the car. These features are actual lifesavers when in unfortunate circumstances the vehicle is subjected to a crash. Many crash tests are carried out on a regular basis to improve the passive safety system that will reduce the fatality rates even after the crash.

When a crash has occurred, then in that instance protection is provided to the occupants from further injury through various features, and these features are called as passive safety features. The protection of the driver and passengers from several crash forces is ensured by the passive safety features. Life space is the protected area surrounding the occupants of the vehicle in which likelihoods of dodging with negligible injuries are possible. Passive safety features help to ensure that the life space exists safe and secure along with the occupants and they remain in this space during the crash.

Fig. 7 Dashboard warning signs

Airbags

Airbags are considered one of the most important and key safety features nowadays. These airbags inflate into cushioning bags essentially saving the driver and the passengers from the crash and from sustaining fatal injuries. Advanced vehicles come with multiple airbags to prevent injuries and to provide maximum protection. Even though airbags do not offer complete protection, it reduces the impact in such a way that the person has a better surviving chance.

Seat Belts

Seat belts along with the pretensioners prevent the person from moving forward suddenly due to inertia when met with a forward collision. This not only prevents the person from being ejected forward but also holds them in a position for the airbags to operate effectively for the safety. Wearing seat belts is mandatory for driving in almost every country, and there are warning lights to indicate if the driver or the driver side passenger doesn't wear it. Modern vehicles include seat belts for the passengers in the rear seats too.

Sturdy Front Design

Another important design feature is a sturdy and rigid front design which will absorb most of the impact on the collision, so that the person inside the vehicle can have better chance of surviving. Careful placing of mounting brackets and other metal or sharp construction away from the areas that are potential to hit the passengers is preferred that in the off chance when a crash is inevitable. Preference is given to leather or soft material usage in areas like dashboard, instrument panel, and other areas of contact.

Non-shattering Windshields

The usage of laminated windshields enhances the visibility, but also the glass comes with a non-shattering feature which breaks into minute particles on crash, thereby avoid breaking into sharp pieces. This in turn reduces the fatal injuries incurred during the crash.

Pedestrian Protective Soft Bumpers

This feature is to prevent sustaining injuries to the pedestrians who may accidentally get hit by the vehicle. Hard front bumpers can inflict severe injuries to the person who gets accidentally hit by the cars. Therefore, relatively softer bumpers are preferred. Redesigning of bumpers, hoods, and windshield to be energy absorbing will actually reduce the impact on pedestrians during pedestrian-car crashes.

Collapsible Steering Column

Steering columns are found to be fatal in sustaining injuries to the driver as it is rigid and immediately contacts the driver when in crash. Therefore, a modification in the design has been brought in where the columns automatically collapse down on high impact, thereby preventing the driver from engaging with a rigid column on crashing.

Cargo Protection

Many SUVs, AUVs, and other utility vehicles have the feature to securely fit the cargo, avoiding it from throwing off or moving while on crash. This also helps from people suffering any injuries from heavy cargo or luggage on the road.

Security

Automotive industry is inching towards superior integration and virtualization by escalating the number of functions and complexity in terms of the software. The reason for this inclination is that the wiring harness has reached its upper boundaries to accommodate ECUs as cars incorporate nearly 100 ECUs nowadays. This results in a broad spectrum of digital and electronic attack surface available where contact is present with various built-in systems along with numerous widely ranging external networks such as Wi-Fi and cellular networks.

A collaborative effort between the supply chain and the comprehensive ecosystem along with a holistic approach is required where their involvement and contribution matters highly to attain security in the complex systems specified above. To attain effective security, it is very important to deal with the components and threats along with the attack points as a whole entity instead of considering them individually. In the context of security, both physical and the cyber worlds should be considered to be compromising as any one of them may lead to devastating results

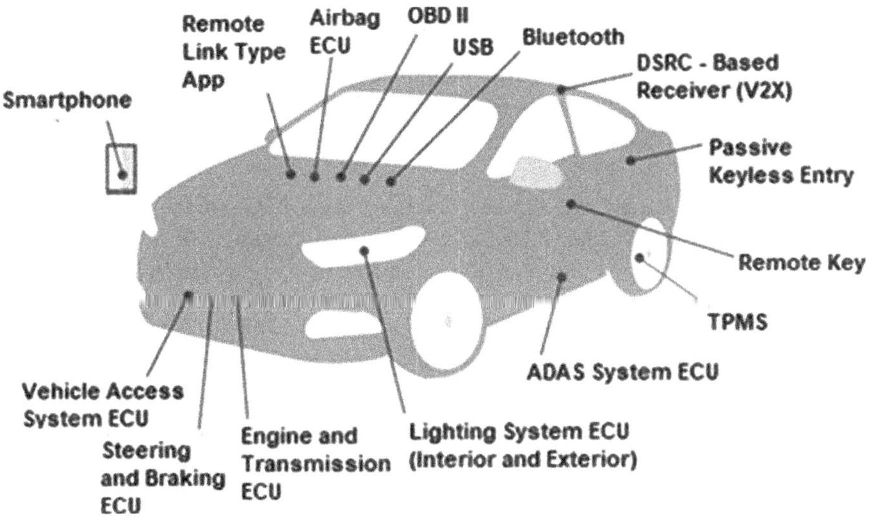

Fig. 8 Surfaces prone to attack due to exposure and hacking on a next-generation car

collectively to a vehicle's systems. The surfaces prone to attack due to exposure and hacking on a next-generation car are illustrated in Fig. 8.

Layers of Security

The three layers that constitute the security in an automobile are, namely, hardware security, network security, and software security.

Hardware Security

The hardware security systems offer physical protection in a car such as the firewall to the engine, seat belts, and airbags. They offer protection to the operating components in case of accidental or intentional damage. Various building blocks of hardware security are used in wide ranges to secure the ECUs and the buses. A few of these hardware security features are described here.

Secure Boot Along with Software Attestation Product keys and digital signatures are checked, thereby detecting tampering if any in the boot loaders as well as critical operating system files. The invalid or tampered files are blocked from execution, providing a trusted foundation on operating ECU, much before it gets infected from tampered files.

Trusted Execution Module A unique identifier is created for each approved component by using cryptographic techniques. This enables an accurate comparison between the known authorized source and a new startup environment. Any attempt made by unknown or unauthorized component or source is prevented as the launch of code is arrested when the match is not found.

Protection from Tampering The user data such as account credentials and intellectual property along with encryption keys are encrypted at compile time. Decryption is performed during a small execution window; thus protection of information is ensured from attempts of tampering and reverse engineering.

Cryptographic Performance Acceleration Cryptographic performance is improved by utilization of optimized hardware for encryption workloads. This ensures easier incorporation of public key or symmetric encryption into various applications and processes. The hardware is designed specifically to enhance and improve the computationally intensive operations involved in encryption.

Protection of Active Memory Instances of buffer overflow need to be prevented as malicious code may exploit them. This is accomplished by embedding the functionality of pointer checking in the hardware; thus vulnerabilities in the code are reduced.

Unique Device Identity Unique identity of each and every device is known to the manufacturers, and it enables secure identification. It prevents unapproved devices from accessing the network and systems intended for a particular manufacturer.

The following sections describe the instances where the hardware security is realized real-time in an automobile.

Smart Keys

Smart keys are electronic entities intended to provide access and authorization to automobiles. Vehicles provided with smart keys can be started without insertion of the key in the ignition as long as the presence of the smart key is ensured within the car. Settings can be adjusted as per the requirements of the users on locking methods such as the press of a button on the door handle or the smart key or the touch on a capacitive area on the door handle. Trunk release is also automated by smart keys electronically along with buttons inside the car which can be used to release the trunk mechanically.

Anti-Theft Alarms

Anti-theft alarms are systems that are installed by manufacturers to prevent unauthorized access of the vehicle. Various systems coupled with GSM and GPS are available where the users are intimated through a message regarding the unauthorized access if done, and based on the cost of the system, the complexity of the system varies. A low-end system sends a message to the user when an unauthorized attempt to steal a car is made, whereas a high-end system files a complaint to the police along with deactivating the fuel injector, thus making the attempt futile.

Software Security

Only direct physical contact of the hackers with the car made the automotive networks vulnerable previously. But nowadays time and money along with unwavering effort from an attacker can yield access to these systems easily without any physical presence too. Common protocols are used to link a huge number of ECUs, and this has potentially amplified the attack surface, and accessibility to vehicles has also increased. ECUs with diverse competencies are present in a vehicle [7]. This increases the difficulty to add hardware security capabilities to some of these ECUs; hence software-based security along with cooperating processors are required. Some of these features that defend the vehicle are described below.

Secure Booting Secure booting cooperates with the hardware and makes sure that the software components that are loaded are valid. This validity ensures for the rest of the system that there exists a root of trust.

Partitioned Operating Systems This involves a general combination of hardware and software intended for isolation of various processes or functions, such as those on the extreme background rather than those that are vital for the vehicle to run in that instant. Hence, the complexity involved in associating multiple systems onto a single ECU is thus decreased. Various techniques such as containers and virtualization aid in assistance for update or replacement of individual functions, preventing any impact on the overall operation [8].

Authentication Authentication provided by just a smart key is no longer enough in this era where never ending advancements in technology are present. Authorization of electronic keys, passwords, and biometrics should be mandatory to access personal information stored in the software of a vehicle as a profile. In the same fashion, authentication by several ECUs is required in order to prevent an attacker gaining control of the system by sending messages or faulty commands.

Imposition of Appropriate Behavior Cyberattacks usually involve compromising one component in a system and using it as a leverage to compromise the rest of the system. Prevention of such an activity in a network is vital for detection and correction of various threats of malicious nature. These are enormously useful and can be utilized in a connected car environment with a collection of cars in the vicinity. Tampering or hacking into one of the cars in the vicinity may pave a way for a cumulative attack by using the affected car as a trusted source for other cars, and this is efficiently evaded.

Advanced Keyless Entry

Advanced keyless entry means that the key or the smart key need not be pressed to unlock or start the car. Just the presence of the key within the required vicinity of the car is enough to unlock the car. The locking and unlocking methods may differ for different vehicles [9].

Network Security

In-vehicle networks contain a combination of operational and personally recognizable information. This may involve navigation routes, current location, call history, etc. It is vital to establish operational security, privacy, and consumer trust, and this is attained by protection of data and messages over the communication bus. Various common protocols such as controller area network (CAN), media-oriented systems transport (MOST), local interconnect network (LIN), automotive Ethernet, FlexRay, Bluetooth, Wi-Fi, and mobile 5G make it possible to protect and safeguard the network since good security techniques that are well tested over the time exist in them. But when new protocols are considered such as dedicated short-range communications (DSRC), they generally magnify the threats, thus intensifying the attack vectors. Security-enhanced ECUs may be coupled with security-enhanced networking protocols, thus ensuring enhanced authenticity, reliability, and integrity of the transmitted data. Without significant hindrance in performance, latency, or real-time response, the following features ensure secure networks [10].

Authentication of the Device and Messages Communications are verified such that they arrive from an approved source. Protection of the authentications is attained so that it is not spoofed or recorded or replayed.

Access Controls Pre-approval of systems and sensors is done, and communications between only the above two are allowed, and the rest of the unapproved messages are blocked along with the alerting regarding any invalid attempts to the security systems. Different access rights are provided to various parties such as manufacturers, end users, drivers, and law enforcement officials to the information systems of the car that have to be authorized and accordingly controlled.

Vehicle Tracking Systems

Vehicle tracking systems are intended to collect the automatic vehicle location of various vehicles and project this massive data into a comprehensive visual representation. This information is used by manufacturers where in case of an accident assistance is provided in some instances. Various popular uses of vehicle tracking systems are theft prevention, stolen vehicle recovery, tracking assets during transportation, fuel monitoring, etc.

Remote Control and Access

In the era of connected vehicles, remote control and access plays a significant role for various tasks like service through remote access, driverless cars controlled from remote operation centers, etc. Latency and system outages are a few hindrances that prevent the remote control and access to be utilized to their full potential in automotive systems. With the advancement of technology with each passing day, these bottlenecks are miniscule. Thus the remote control has various energy and environmental impacts where drive cycle is optimized which in turn improves the fuel economy.

Summary

The various vehicular safety features prevalent in the automobiles to enhance the safety as well as security are explained here. The escalating growth in the field of vehicular safety is motivated by a combination of dynamic entities. In the era of connected vehicles, connectivity of the vehicles with the external world is essential. Due to this the attack surface is extremely widened leading to vulnerabilities. Compromising connectivity to address the vulnerabilities is not a solution; thus, advanced safety and security measures should be embedded in the vehicles. The vehicles should be well equipped to handle such attacks once in a while, and such sophisticated safety and security maneuvers are considered mandatory by the automakers nowadays.

References

1. X. Wang, Y. Jiang, R. Li, M. Chen, Automobile safety technology and its improvement, in *MATEC Web of Conferences*, vol. 160, (2018), p. 05012. https://doi.org/10.1051/matecconf/201816005012

2. L. Moravčík, M. Jaśkiewicz, Boosting car safety in the EU, in *2018 XI International Science-Technical Conference Automotive Safety, Casta*, (2018), pp. 1–5. https://doi.org/10.1109/AUTOSAFE.2018.8373307
3. B. Deng, X. Zhang, Car networking application in vehicle safety, in *2014 IEEE Workshop on Advanced Research and Technology in Industry Applications (WARTIA)*, (Ottawa, ON, 2014), pp. 834–837. https://doi.org/10.1109/WARTIA.2014.6976402
4. C. Li, J. Wang, H. Li, X. Liu, Y. Wang, Intelligent control system of automobile front-light based on active safety, in *2017 9th International Conference on Measuring Technology and Mechatronics Automation (ICMTMA)*, (Changsha, 2017), pp. 133–136. https://doi.org/10.1109/ICMTMA.2017.0040
5. M. Vyas, H. Sarath, K. Smitha, A. Bagubali, A strategy and framework for analysis of operational data of automotive radars for development of active safety systems, in *2017 2nd IEEE International Conference on Recent Trends in Electronics, Information & Communication Technology (RTEICT)*, (Bangalore, 2017), pp. 2176–2181. https://doi.org/10.1109/RTEICT.2017.8256986
6. H. Ludanek, S. Patuschka, Improved safety and security through vehicle electronics, in *2006 International Conference on Applied Electronics*, (Pilsen, 2006), pp. 1–1. https://doi.org/10.1109/AE.2006.4382948
7. Automotive Security Best Practices. Recommendations for security and privacy in the era of the next-generation car. McAfee, San Jose.
8. A. Atamli-Reineh, A. Martin, Securing application with software partitioning: A case study using SGX, in *Security and Privacy in Communication Networks. SecureComm 2015. Lecture Notes of the Institute for Computer Sciences, Social Informatics and Telecommunications Engineering*, ed. by B. Thuraisingham, X. Wang, V. Yegneswaran, vol. 164, (Springer, Cham, 2015). https://doi.org/10.1007/978-3-319-28865-9_40
9. N.S. Bhuvaneswari, M. Raghavan, Intelligent safety and security systems in automobiles, in *2015 IEEE Technological Innovation in ICT for Agriculture and Rural Development (TIAR)*, (Chennai, 2015), pp. 188–192. https://doi.org/10.1109/TIAR.2015.7358555
10. R. Soja, *Automotive Security: From Standards to Implementation* (Freescale, NXP)

Virtualizing an Automotive State-of-the-Art Microcontroller: Techniques and Its Evaluation

Arun Kumar Sundar Rajan and M. Nirmala Devi

Introduction

Cars are no longer just vehicles of transportation but a complex and innovative electronic entity which ensures our safety and has become an integral part of our day to day life. Over the last two decades, technological innovations in the automotive domain started replacing the traditional mechanical components with more and more electrical and electronic devices. Luxury cars like BMW 7 series have more than 150 electronic control units (ECUs) [1]. Though these devices facilitate modularity in operation, it increases the cost of deployment and maintenance. Original equipment manufacturers (OEMs) have taken an approach of restructuring the E/E architectural components into specific domains or zones [2], with a future focus of achieving a centralized computing platform (CCP) [3]. Consolidation of small applications like acoustic vehicle alerting system (AVAS), ventilation system (HVAC), and charge control unit (CCU) into the zone of vehicle control unit (VCU) is an example of such restructuring process.

For quite some years, hypervisor (HV) had been a known solution for consolidating real-time heterogeneous applications onto the same hardware and at the same time ensuring isolation of temporal, spatial, and faults among the applications. When a system is virtualized, it is possible to plug and play an application without interfering other existing applications [4]. Efficiency and performance of the hypervisor primarily depends on how the underlying hardware can support for virtualization.

Advanced processors with built-in virtualization features are used in automotive domain for zones like infotainment, but not yet deployed in safety critical zones like

A. K. Sundar Rajan (✉) · M. Nirmala Devi
Department of Electronics and Communication Engineering, Amrita School of Engineering, Amrita Vishwa Vidyapeetham, Coimbatore, India
e-mail: s_arunkumar@cb.amrita.edu; m_nirmala@cb.amrita.edu

M. Kathiresh, R. Neelaveni (eds.), *Automotive Embedded Systems*, EAI/Springer Innovations in Communication and Computing,
https://doi.org/10.1007/978-3-030-59897-6_2

powertrain and vehicle control functions. Compared to controller chips [5], processors generally do not own the data they process nor control, and they depend on the reliability of external memory and its data transfer. This brings an uncertainty on the safety and security aspect! Additionally, execution of ASIL-D [6] powertrain functions must be monitored and verified by the hardware—where MCUs with *lockstep core* functionality suffice this requirement.

In view of safety and security standards, automotive original equipment manufacturers (OEMs) still insist on pursuing the powertrain functionality development on multicore **controller** platform (MCU) instead of advanced many-core **processors**. Although automotive MCUs have many cores and execute at higher clock speeds, the common problem is that these MCUs do not facilitate hardware-assisted extensions for virtualization. Furthermore, hardware resources and response time of MCUs are still a bottleneck compared to that of a personal computer (PC). It is also critical in a virtualized system to share the MCU's hardware resources like memory, peripherals, and system clock among the applications or virtual machines (VMs).

In this chapter, the following techniques for virtualizing a legacy system are discussed:

- Securing a virtual machine
- Synchronizing the start-up of a virtualized system
- Handling of hardware interrupts and traps
- Facilitating peripheral access (input/output access) to the applications

A demonstrator with two heterogeneous virtual machines, namely, application for governing the engine control unit (E-ECU) and vehicle control unit (VCU), was used. These applications were ported to a real-time automotive powertrain controller (TC29x-AURIX) [7], and a prototype version of the hypervisor from ETAS GmbH [8] was used. Detailed findings from our demonstrator are presented in the reference [9]. Experimental observations from the demonstrator are put forth, followed by recommendations and limitations of the virtualized system.

Hypervisor Classification

Hypervisor, also termed as virtual machine monitor (VMM), is a software technique to consolidate multiple virtual machines (VM) or applications on the same hardware. A common example, most of the users would have Windows OS installed on their laptop. Now, if the user wanted to use the same laptop to work with LINUX OS—there are two options:

- To uninstall windows and install LINUX
- Inside the Windows environment, the user shall install an application called "VMware," and install LINUX inside the VMware

The former option restricts the user from accessing applications in Windows. However, the latter supports the user to access applications of windows and LINUX simultancously on thc samc laptop. Hcrc, VMwarc plays thc rolc of hypcrvisor/VMM, enabling applications running on Windows and Linux to use the same, underlying the laptop's memory, peripheral devices, etc.

Having the context of hypervisor set, the next step is to classify based on the deployment [10, 11] of hypervisor.

Type 1 or Baremetal Hypervisor

Hypervisor is a software layer that executes directly on the hardware, encapsulating all the access towards the hardware. The operating system (OS) is part of the application, referred as *guest OS*. Applications along with their OS work like a plug-and-play components, as shown in Fig. 1a. The hypervisor layer forwards the hardware interrupts to the application and also serves the request for hardware access.

Type 2 or Hosted Hypervisor

In contrast to the type 1 hypervisor, the hardware comes with an OS (host OS), and the hypervisor is implemented on top of the host OS. In this type, the hypervisor cannot directly access the hardware and instead use the services provided by the host OS, as shown in Fig. 1b. The applications can also have their own OS (guest OS). Compatibility between host OS and hypervisor is a decisive point for the design.

It has been over several decades since the hypervisor has found its use in servers deployed at data centers. A brief history on the journey of virtualization is presented in works of Strobl et al. [12]. An embedded environment is a miniature of PC environment, designed for a specific purpose with limited hardware resources but robust.

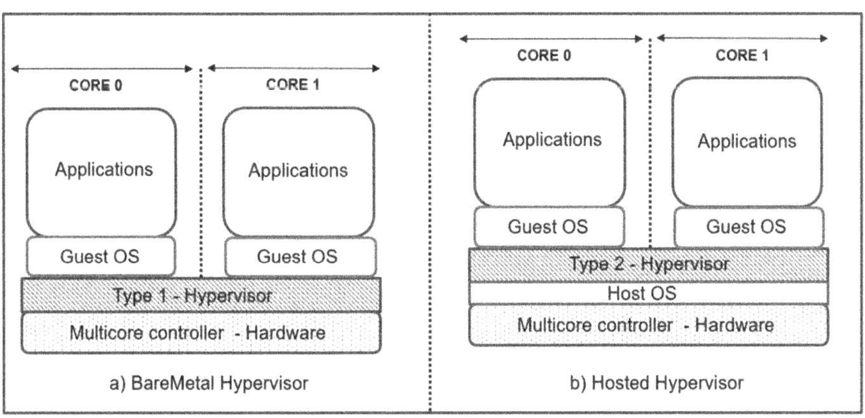

Fig. 1 Overview of type 1 and type 2 hypervisor

Real-time embedded systems have always been hindered from virtualization, because of hard real-time deadlines and limited resources to share [13, 14]. With introduction of fast multicore embedded chips [15], virtualization seems to be a possible attempt [12]. There are different types of embedded hypervisors, commercial as well as open source. A detailed comparison of four known embedded hypervisors is presented in our previous work [9].

Implementation

In the automotive domain, the target is not just to execute an application but to ensure they are executed in hard real-time and fail-safe. Considering the safety and security standards that are to be fulfilled, automotive original equipment manufacturers (OEMs) pursue the powertrain functional development on multicore **controller** platform (MCU) instead of advanced many-core **processors**. Compared to controller chips, processors generally do not own the data they process, and they depend on the reliability of external memory and its data transfer.

Step 1. Know About Your Hardware

Some of the known automotive state-of-the-art MCUs are as follows:

- Hercules TMS570—ARM Cortex-R4 architecture [16]
- Infineon's AURIX—TriCore architecture [17]
- Freescale's MPC5643L MCU—Power architecture [18]
- Renesas' H50/P1x MCU [19]
- STMicroelectronics' SPC5 MCU—Power architecture [20]

The level of safety and security aspects is one step ahead in these devices, as they ensure at the hardware level if the executed instruction is correctly handled or not (*lockstep cores*). A common downside is that these MCUs do not possess the instruction set architecture (ISA) [21] to support virtualization. For controllers lacking *hypervisor* mode, a different strategy is required to implement virtualization. In this section, details about virtualizing such a MCU (AURIX TC29x) are elaborated.

Infineon's AURIX family TC29x microcontroller is an automotive state-of-the-art multicore controller, with three independent 32-bit TriCore™ CPUs (also termed as *core*). A block diagram of the AURIX TC29x controller is shown in Fig. 2 [9]. A user guide and details of microcontroller are available in the reference webpage [22].

An important strategy for virtualizing a hardware is to know about the modes of access control over its memory and peripheral regions. MCUs with hypervisor-assisted extensions possess a separate operating mode with limited access to the hardware. In the case of AURIX TC29x controller, there is no hypervisor mode. It

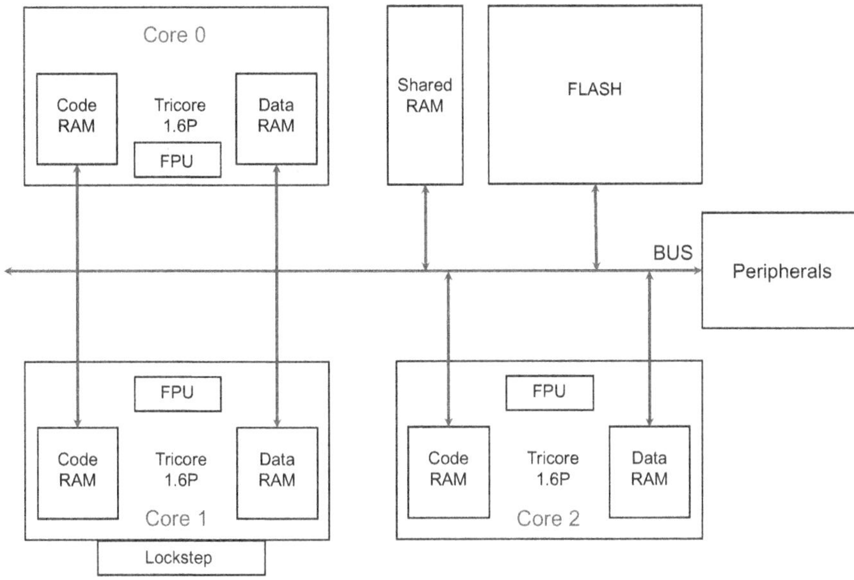

Fig. 2 Block diagram of AURIX TC29x controller [9]

has a memory protection unit (MPU) and three standard operating modes, namely, user-0, user-1, and supervisor mode [22]:

User-0 Mode: This mode does not allow access to any peripheral devices. Enabling or disabling of interrupts is also not permitted.

User-1 Mode: This mode allows access to unprotected peripherals like read/write access to serial port, read access to timer, and I/O status registers. It also allows disabling interrupts for a shorter period, which can be overridden by system control register.

Supervisor Mode: This mode permits read/write access to system registers and all peripheral devices. It also allows enabling and disabling of interrupts.

From the hypervisor perspective, applications run independently. These applications will approach hypervisor for hardware accesses and to communicate with other applications. The strategy for MCUs without hypervisor extension is to exploit a combination of MPU and existing standard operating modes. Based on privileges of the operating modes, applications are made to execute in a lesser privileged *user-1* mode, and only hypervisor executes in *supervisor* mode.

Step 2. Securing the Partitioned Memory Region

The next step is to distribute the heterogeneous applications onto the hardware. Consider a scenario of three heterogeneous applications, with each application distributed to a core of TC29x controller. The same is depicted in Fig. 3. Thus, the

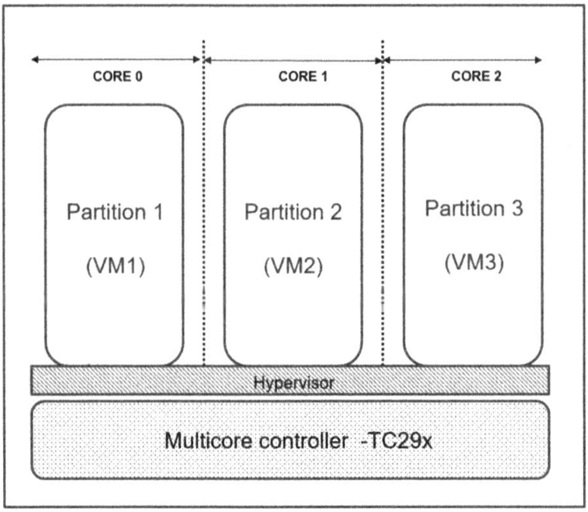

Fig. 3 Distribution of applications in a multicore device

entire device is divided into three virtual partitions. *Note*: There are some commercial hypervisors which are capable of handling multiple VMs on the same core.

The memory and peripheral address region for each partition is governed using MPU. MPU also monitors the access of an application (read, write, and execute). MPU understands the regions to be protected by using its base/limit register pairs, which hold the access type and memory address range details. Each MPU base/limit register pair has to be configured statically. Violation of access to the protected memory region will end up in a trap. For a nonintrusive implementation of a virtualized system, the entire address range in terms of flash, RAM, and peripheral address space is recommended to be protected using MPU.

In additional to MPU configuration, there are few basic configurations needed for each partition. These configurations are done based on the application, and this information is needed for the hypervisor.

(a) Configuration of interrupt handlers and their priority
(b) Configuration of exception/trap handlers

Step 3. Setting Up of the Start-Up Sequence

In a multicore controller, during power-on (or reset), only one core will begin its execution. This core performs boot-up sequence of the controller. A simple explanation of the boot-up sequence can be seen in Jacob's blog [23]. Once the checks are complete, the control is handed over to the entry point (*_main*) of the application. This entry point will commence the start-up sequence of other cores, as depicted in

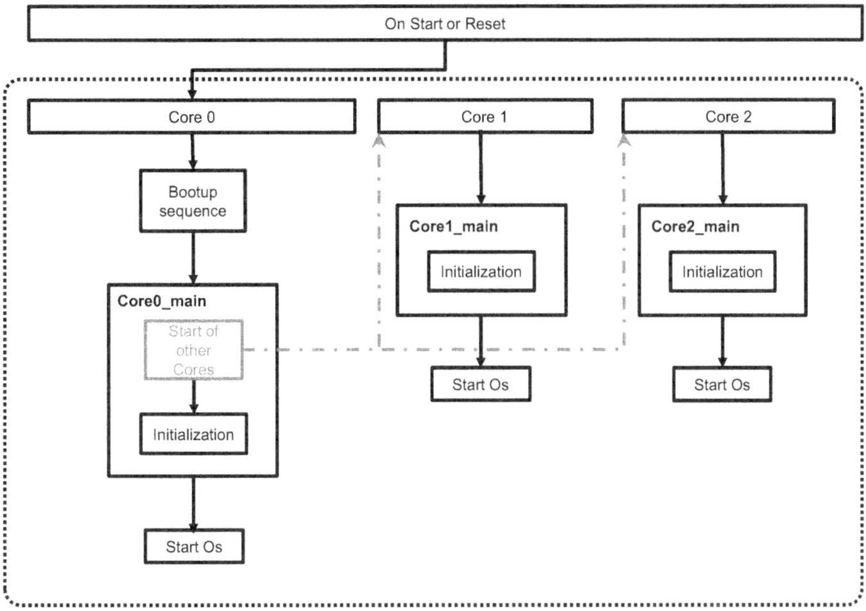

Fig. 4 Start-up sequence of a conventional multicore system

Fig. 4. The start-up sequence will perform initialization and then transfer the control to the operating system of the corresponding application.

The hypervisor should know the number of cores that it will handle, and the hypervisor plays the role of starting other cores. Figure 5 depicts the start-up sequence in a virtualized environment. Hypervisor starts the cores at *supervisor* mode, and after the initialization, the hypervisor switches the execution to *user-1* mode before handling over the control to OS.

Step 4. Handling of Traps and Interrupts

Traps and interrupts are the basic tools of a hardware to inform the executing application about an event occurrence. In turn, the application responds to the interrupt via its ISRs. In the virtualized system, interrupt and/or trap is first handled at the hypervisor layer and not by the applications. The challenge is to invoke the right application's handler on event occurrence.

As described in step 2, each partition or VM will hold a configuration of trap and interrupt handlers. This is a static configuration, done based on the nature of the application and its required peripherals.

A partition's configuration of interrupt/trap would consist of the following information:

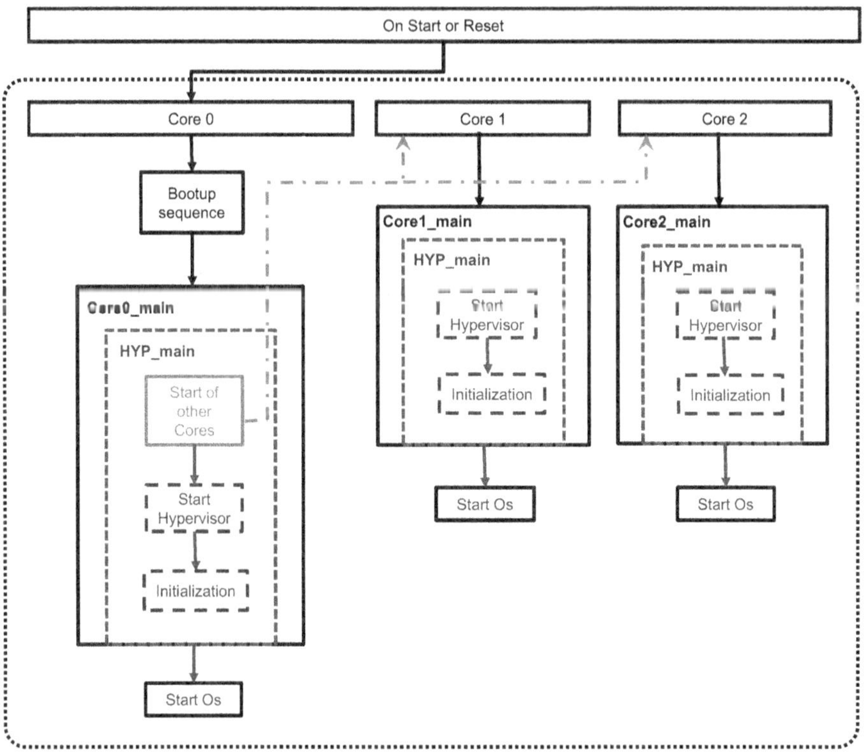

Fig. 5 Start-up sequence of a virtualized system

- Exception vector ID
- Exception table number
- Exception priority (needed only for interrupts)
- Exception action

When a core is divided into multiple partitions or VMs, then for each partition of the core, an exception table will be needed. When there is only one partition per core, as described in this chapter, only one exception table per core will be present.

When an event occurs, there are two possible *exception actions* at the hypervisor layer:

(a) To handle the event at the hypervisor layer itself
(b) To forward it for handling at the corresponding partition

A summary of the possible event handlers at hypervisor layer is presented in Fig. 6. It is possible to emulate the event handlers and perform input–output operation.

If an application executes a *load* or *store* instruction, then the hardware triggers a corresponding trap for *read/write* violation. In the hypervisor layer, a trap handler emulates the *read/write* instruction and returns the result of the emulated

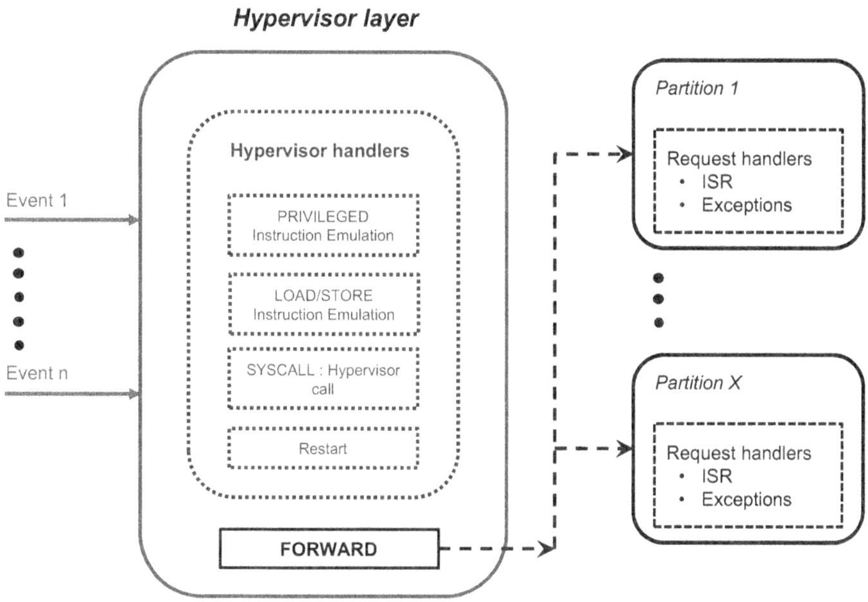

Fig. 6 Event handlers in a virtualized environment

instruction. Emulation of the event handlers can also be extended to handle *privileged* instructions. Default action is to *reset* the execution and begin from the start.

Step 5. Input/Output (I/O) Peripheral Access

Embedded MCUs have limited resources in terms of memory and peripherals compared to that of a personal computer (PC). Some of the commonly seen peripherals in automotive MCUs are ADC, DAC, timers, and communication peripherals for CAN, Ethernet, FlexRay, etc. which are broadly grouped as input–output (IO) peripherals. A summary of various approaches in virtualizing the IO peripheral in order to share between VMs is illustrated in Fig. 7.

Direct Device Assignment

Memory protection unit (MPU) [7] of the hardware is utilized for dedicated assignment of a peripheral to a particular VM. The required peripheral's address range is configured in the MPU register set, and the VM corresponding to that register set will get direct access to the peripheral. Another advantage is that VM can now access the peripheral directly in *user* mode itself. If any another VM tries to access

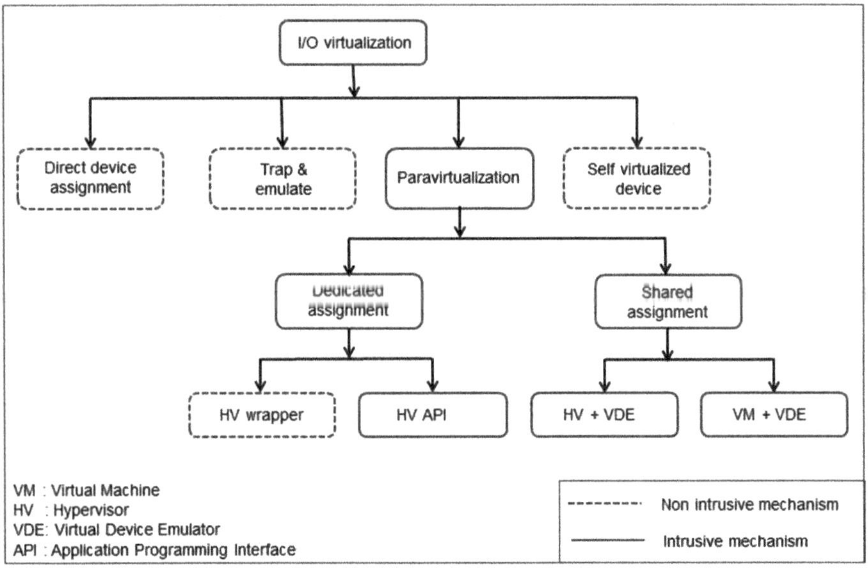

Fig. 7 Various methods of IO Virtualization

this peripheral, a trap will be triggered by the hardware. It is a nonintrusive mechanism. A drawback is the limitation in the number of MPU register sets provided by the hardware which may not be sufficient to govern the complete peripheral address range, as illustrated in the reference [9].

Trap and Emulate

It is a nonintrusive mechanism. When a VM executes a privileged instruction or tries to access the peripheral address range in user mode, a trap is triggered, and the control enters a trap handler. As discussed in step 4, the trap handler is emulated inside HV. The trap handler validates the request triggered by VM and emulates the functional behavior of the requested instruction.

Sugerman et al. [24] handled privileged instructions using a trap and emulate mechanism, where a significant overhead in terms of execution time is presented. In our previous work [9], we have utilized this mechanism for emulating IO access and for privileged instruction handling. To summarize, drawbacks of this method are the need for exhaustive instruction set emulation and the timing overhead caused by trap and its handler emulation.

Self-Virtualized Device

This technique is also called as hardware-assisted virtualization [25], where the peripheral handling is taken care of by a dedicated peripheral hardware. Herber et al. [26] used HW-assisted virtualization with direct device access on Intel Core i7-3770T. Their results show hardware-assisted virtualization can offer lower latency and better predictability compared to paravirtualization. However, their implementation [27] concludes with the limitation of incorporating hardware virtualization in automotive multicore controllers (based on AURIX controller TC27x). This technique also requires interfacing to external ASICs, where the data reliability for ASIL-D has to be ensured again. Another aspect that limits this technique is the availability of virtualized hardware for all the required peripherals.

Paravirtualization

Paravirtualization is an intrusive mechanism which requires modification in the application's software in order to integrate it into a virtualized environment. As depicted in Fig. 7, there are four possible ways to classify IO paravirtualization depending on dedicated or shared peripheral usage. In dedicated peripheral usage, a VM can own its peripheral address region and access it via wrapper or API techniques.

- *HV wrapper*: A replace function is invoked automatically when VM executes an instruction to accesses its peripheral. Replace function executes from HV space in *supervisor* mode [9]. Switching from user mode to supervisor mode is handled implicitly; this replace function accesses the peripheral and returns the required result.
- *HV API*: VM has to use the library APIs provided by HV vendor in order to access the peripheral. Here, VMs have to be adapted to integrate HV API calls [9]. The type of access depends on the library provided by HV vendor.

Shared paravirtualization is complex compared to the above approach. It requires additional control of routing between appropriate VM and hardware, arbitration among VMs, and data consistency check during concurrent access. A *virtual device emulator* (VDE) is a functional block which emulates the peripheral's device driver functions with supervisor access rights. Arbitration and synchronizing algorithms [28] to handle concurrent access from multiple VMs are implemented inside VDE. Based on how the VDE is interfaced with VMs, shared paravirtualization is further classified as follows:

- *VM + VDE*: Device emulator is placed within a dedicated master VM. Other slave VMs that need access to the peripheral communicates via this master VM. Herber et al. [29] and Reinhardt et al. [30] prototyped paravirtualization using an Infineon AURIX TriCore TC27x. They have integrated all drivers in an independent virtual machine and routed all hardware access over that central virtual instance. Frontend and backend communication model was used. Gabriela et al. [31] have also attempted a similar approach on Multi-Processor System-on-Chip (MPSoC), where the master VM has access to the hardware, while other VMs have to communicate to the hardware via this master VM. Menon et al. [32] profiled XEN hypervisor, which showed a throughput degradation of 20%, when the driver domain and VM are executed on the different cores, and up to 66% degradation if they share the same core. The overhead is mainly constituted by data copy operations and context switches between guest and host mode. To summarize, disadvantages of this technique are the timely availability of the master VM to respond on slave VM's request, overhead, and reliability of inter-VM communication model.

- *HV + VDE*: Device emulator for the peripheral is deployed inside the hypervisor layer. VMs access the peripheral by calling VDE API, and VDE communicates back to the corresponding VM using either interrupts or shared memory. Advantage of this approach is each VM can directly communicate with the peripheral independent of other VMs.

Experimental Observations

A demonstrator with two heterogeneous virtual machines, namely, application for governing the engine control unit (E-ECU) and vehicle control unit (VCU), was used. These applications were ported to a real-time automotive powertrain controller (TC29x-AURIX) [7], and a prototype version of the hypervisor from ETAS GmbH [8] was used. Detailed findings from our demonstrator are presented in the reference [9].

Memory Allocation Plan

Flash and RAM regions have to be evaluated based on code and data consumption. This evaluation helps in partitioning the memory regions and securing them using MPU. (*Refer to step 2 of implementation.*)

(a) Evaluate the amount of memory needed for application.
(b) Evaluate the amount of memory needed for hypervisor module.
(c) Plan additional buffer space to accommodate changes based on hypervisor configuration.
(d) Identify dedicated peripherals and configure their address range under MPU.

Table 1 Overhead with hypervisor codes [9]

Functions	Execution time (μs)
Function that checks for memory range	3.1–3.3
Emulation function for input–output access	2.09–2.93
Emulation function for privilege instruction: MTCR	2.33
Emulation function for privilege instruction: BISR	1.36
Hypervisor main function	8.15

Execution Time

Overhead from Hypervisor Function

In the section of implementation, various functions of hypervisor are discussed. Execution of the hypervisor functions is considered as overhead, because these functions are not part of the required application. Table 1 is an extract from our detailed work presented in reference [9]. Execution time [33] varies depending on the type of emulated instruction. *For example, an absolute addressing consumes less instructions than offset addressing.*

Core Load

The amount of computational work handled by a core in a given cycle is an important measure for evaluating the load balancing algorithm and performance of the overall virtualized system.

Key points from our experiment [9] are as follows:

(a) Application's direct access to MPU configured address ranges will not increase the CPU load.
(b) Emulation of privileged instructions will account for 8–10% increase in CPU load.
(c) The entire peripheral sharing via IO emulation technique can stall the execution of a core. This high load situation can occur when there are simultaneous requests for emulation of load/store instructions and trigger of frequent hardware interrupts.

Influence on Interrupt Timing

ADC and CAN interrupts are evaluated with the demonstrator [9]. An interrupt reaches an application after going through the hypervisor layer. The major contributors for the increased latency in interrupt handling are as follows:

(a) Decision-making at the hypervisor layer, instead of direct forward of interrupt request to the application
(b) Switching time between user-1 mode and supervisor mode
(c) Partition's interrupt service routine (ISR) executing in user 1 mode, when any privileged instruction will go through an additional emulation

Behavior of Input–Output Access

In order to evaluate the real-time behavior of peripheral access, a CAN node is virtually shared between the applications (E-ECU and VCU) of the demonstrator. Technique of HV + VDE is exploited. Periodic CAN frames are transmitted at specified time intervals, whose periodicity depends on the criticality and rate of change of information carried by the frame. The quickest periodic request that can be made from the demonstrator has an interval of 10 ms. The evaluation of the response time under simultaneous requests to the same CAN node at various combinations of periodicity is presented in Table 2.

From Table 2, a deviation of 1–3 ms is seen when both VMs try to access the same CAN node with the *same CAN message ID*. In an automotive CAN network design, a particular CAN message ID will have **a unique sender** and multiple

Table 2 Response time of periodic CAN communication

Scenario	Periodic transmission frame request from E-ECU (ms)	Periodic transmission frame request from VCU (ms)	Response time (range of values)
1	50	50	E-ECU: 49.879–52.078 ms
			VCU: 49.964–51.049 ms
2	50	10	E-ECU: 49.782–52.158 ms
			VCU: 9.998–12.008 ms
3	20	10	E-ECU: 19.995–21.004 ms
			VCU: 9.997–12.610 ms
4	10	50	E-ECU: 9.857–11.165 ms
			VCU: 49.967–51.031 ms
5	10	10	E-ECU: 9.977–13.021 ms
			VCU: 9.999–13.009 ms

receivers. However, we simulated a rare scenario to have two senders for the same CAN message ID in order to see the delay. With usage of the same CAN message ID, prioritization during HV arbitration is avoided.

Also this time deviation happens only when both VMs place the request simultaneously. *For example: in the fastest scenario of 10 ms, one of the VMs have to face a delay of 3 ms.* In most of the automotive use cases, a deviation of up to ±1 ms is acceptable.

Shortcomings and Recommendations

During the evaluation of the demonstrator [9], some of the critical observation points are summarized in this subsection.

Cost of Virtualization

To realize a virtualized environment, many existing legacy applications will be squeezed onto a single multicore hardware. In such an attempt, the following cost factors must be accounted:

1. Adaptations performed on a legacy application to port onto the desired hardware
2. Modifying the application's hardware access via a hypervisor API call
3. Configuration effort of each partition under hypervisor
4. Additional performance and runtime delay from the hypervisor overheads

Boundary Constraints

When virtualizing an embedded real-time system, there are certain boundaries, namely:

1. For data protection and freedom from interference, applications have to execute in lesser privileged mode, without any direct access to hardware.
2. Only the hypervisor functions execute with higher privileges, with complete access to hardware. Hence, applications have to rely on the hypervisor for performing many basic functions.
3. As stated in step 4, an interrupt to an application will first go to the vector table of the hypervisor and then be forwarded, which obviously increases the interrupt's response time.
4. Access to shared resources has to go through a hypervisor's arbitration protocol, and this will slow down the access.

5. Memory management is solely responsible to ensure spatial protection. Micro assignment of hardware peripherals is bounded by the extensiveness of memory management.

Recommendations

Recommendations for setting up a virtualized system are as follows:

1. Interrupt service routine for each application should be handled locally in their corresponding core. Centralized interrupt handlers will only bring in additional delay and overload to that centralized CPU.
2. In order to reduce interrupt response time, a centralized interrupt vector table pointing to interrupt handlers of corresponding application should be considered. However, this technique will not be applicable to shared peripheral and its interrupt.
3. Microcontroller abstraction layer (MCAL), which is specific to hardware, should be integrated into the hypervisor layer.
4. Applications should be independent of hardware, which eases portability and interfacing with the hypervisor.
5. Hardware with virtualization extensions like *hypervisor mode* and with advanced memory management unit (MMU) [34] will reduce the overall delays and overhead.

Conclusion

Our work is a demonstration of the feasibility to virtualize an automotive state-of-the art microcontroller, which does not possess any ISA for virtualization. From our demonstrator [9], we were able to convey the techniques and downside of the virtualization. There are many parts inside this implementation that can be optimized to earn better response timing such as arbitration handling algorithm and periodicity at which the software is executed; usage of hardware with virtualization features and multiple vector table support can help to directly forward the requests to a corresponding application.

Virtualization is a prospective for consolidating heterogeneous application onto a single efficient hardware, at the same time ensuring isolation from interference. Additional overheads like timing delays introduced with hypervisor and its algorithm have to be accounted during the system design.

References

1. BMW car models, April 2019. [Online]. Available: https://www.bmw.in/en/all-models/7-series/sedan/2015/equipment.html
2. D. Reinhardt, M. Kucera, Domain controlled architecture, in *Proc. Third International Conference on Pervasive and Embedded Computing and Communication Systems (PECCS 2013)*, (2013)
3. G. Weiss, P. Schleiss, C. Drabek, Towards flexible and dependable E/E-architectures for future vehicles, in *4th International Workshop on Critical Automotive Applications: Robustness & Safety (CARS 2016)*, (2016)
4. A. Crespo, M. Masmano, J. Coronel, S. Peiró, P. Balbastre, J. Simo, Multicore partitioned systems based on hypervisor. IFAC Proc. **47**(3), 12293–12298 (2014)
5. Difference between microprocessor and microcontroller, electronicsforu, April 2019. [Online]. Available: https://wwww.electronicsforu.com/resources/difference-between-microprocessor-and-microcontroller
6. Y. Gheraibia, S. Kabir, K. Djafri, H. Krimou, An overview of the approaches for automotive safety integrity levels allocation. J. Fail. Anal. Prev. **18**(3), 707–720 (2018)
7. AURIX™ Family – TC29xT, Infineon AG, April 2019. [Online]. Available: https://www.infineon.com/dgdl/Infineon-TC29x_B-step-UM-v01_03-EN.pdf?fileId=5546d46269bda8df0169ca1bdee424a2
8. D. Reinhardt, G. Morgan, An embedded hypervisor for safety-relevant automotive E/E--systems, in *Proceedings of the 9th IEEE International Symposium on Industrial Embedded Systems (SIES 2014)*, (IEEE, 2014), pp. 189–198
9. A.K.S. Rajan, A. Feucht, L. Gamer, I. Smaili, Hypervisor for consolidating real-time automotive control units: Its procedure, implications and hidden pitfalls. J. Syst. Archit. **82**, 37–48 (2018)
10. A. Patel, M. Daftedar, M. Shalan, M.W. El-Kharashi, Embedded hypervisor xvisor: A comparative analysis, in *2015 23rd Euromicro International Conference on Parallel, Distributed, and Network-Based Processing*, (IEEE, 2015), pp. 682–691
11. M. Mounika, C. Chinnaswamy, A comprehensive review on embedded hypervisors. Int. J. Adv. Res. Comp. Eng. Technol. **5**(5) (2016)
12. M. Strobl, M. Kucera, A. Foeldi, T. Waas, N. Balbierer, C. Hilbert, Towards automotive virtualization, in *2013 International Conference on Applied Electronics*, (IEEE, 2013), pp. 1–6
13. H. Sutter, The free lunch is over: A fundamental turn toward concurrency in software. Dr. Dobb's J. **30**(3), 202–210 (2005)
14. J. Savithry, A.G. Ortega, A.S. Pillai, P. Balbastre, A. Crespo, Design of criticality-aware scheduling for advanced driver assistance systems, in *2019 24th IEEE International Conference on Emerging Technologies and Factory Automation (ETFA)*, (IEEE, 2019), pp. 1407–1410
15. S. Abinesh, M. Kathiresh, R. Neelavenik, Analysis of multi-core architecture for automotive applications, in *2014 International Conference on Embedded Systems (ICES)*, (IEEE, 2014), pp. 76–79
16. Texas instruments – Automotive MCU, April 2019. [Online]. Available: http://www.ti.com/lsds/ti/microcontrollers-16-bit-32-bit/c2000-performance/safety/tms570/overview.page.
17. Infineon devices, April 2019. [Online]. Available: https://www.infineon.com/cms/en/product/microcontroller/32-bit-tricore-microcontroller/32-bit-tricore-aurix-tc2xx/aurix-family-tc29xt/?redirId=100842.
18. Freescale automotive MCU, April 2019.. [Online]. Available: https://www.nxp.com/docs/en/data-sheet/MPC5643L.pdf.
19. Renesas automotive MCU, 2019.. [Online]. Available: https://www.renesas.com/en-in/products/microcontrollers-microprocessors/rh850/rh850p1x/rh850p1hc.html.
20. ST automotive MCU, April 2019. [Online]. Available: https://www.st.com/en/automotive-microcontrollers/spc5-32-bit-automotive-mcus.html?querycriteria=productId=SC963.

21. J. Fisher-Ogden, *Hardware support for efficient virtualization* (University of California, San Diego, Tech. Rep, 2006), p. 12
22. AURIX™ Family – TC29xT, Infineon AG, April 2019. [Online]. Available: http://www.infineon.com/cms/en/product/microcontroller/32-bit-tricore-tm-microcontroller/aurix-tm-family/aurix-tm-family-%E2%80%93-tc29xt/channel.html?channel=db3a304342c787030142dc92c9aa1674.
23. Jacob Beningo's understanding-the-microcontroller-boot-process, April 2019. [Online]. Available: https://www.beningo.com/understanding-the-microcontroller-boot-process.
24. J. Sugerman, G. Venkitachalam, B.H. Lim, Virtualizing I/O Devices on VMware workstation's hosted virtual machine monitor, in *USENIX Annual Technical Conference, General Track*, (2001), pp. 1–14
25. P. Varanasi, G. Heiser, Hardware supported virtualization on ARM, in *Proceedings of the Second Asia-Pacific Workshop on Systems*, (2011), pp. 1–5
26. C. Herber, A. Richter, H. Rauchfuss, A. Herkersdorf, Spatial and temporal isolation of virtual can controllers. ACM SIGBED Rev. **11**(2), 19–26 (2014)
27. C. Herber, A. Richter, H. Rauchfuss, A. Herkersdorf, Self-virtualized CAN controller for multi-core processors in real-time applications, in *International Conference on Architecture of Computing Systems*, (Springer, Berlin, Heidelberg, 2013), pp. 244–255
28. M. Belwal, T.S.B. Sudarshan, Intermediate representation for heterogeneous multi-core: A survey, in *2015 International Conference on VLSI Systems, Architecture, Technology and Applications (VLSI-SATA)*, (IEEE, 2015), pp. 1–6
29. C. Herber, D. Reinhardt, A. Richter, A. Herkersdorf, HW/SW trade-offs in I/O virtualization for controller area network, in *2015 52nd ACM/EDAC/IEEE Design Automation Conference (DAC)*, (IEEE, 2015), pp. 1–6
30. D. Reinhardt, M. Güntner, S. Obermeir, Virtualized communication controllers in safety-related automotive embedded systems, in *International Conference on Architecture of Computing Systems*, (Springer, Cham, 2015), pp. 173–185
31. G. Breaban, M. Koedam, S. Stuijk, K. Goossens, Virtualization and emulation of a CAN device on a multi-processor system on chip, in *2016 5th Mediterranean Conference on Embedded Computing (MECO)*, (IEEE, 2016), pp. 41–46
32. A. Menon, J.R. Santos, Y. Turner, G. Janakiraman, W. Zwaenepoel, Diagnosing performance overheads in the xen virtual machine environment, in *Proceedings of the 1st ACM/USENIX International Conference on Virtual Execution Environments*, (2005, June), pp. 13–23
33. A. Kohn, K. Schmidt, J. Decker, M. Sebastian, A. Züpke, A. Herkersdorf, Timing analysis for hypervisor-based I/O virtualization in safety-related automotive systems. SAE Int. J. Passeng. Cars-Electron. Electr. Syst. **10**, 368–379 (2017)
34. C. Moratelli, F. Hessel, Hardware-assisted interrupt delivery optimization for virtualized embedded platforms, in *2015 IEEE International Conference on Electronics, Circuits, and Systems (ICECS)*, (IEEE, 2015), pp. 304–307

AUTOSAR and MISRA Coding Standards

Y. Catherine Yamili and M. Kathiresh

Introduction: AUTOSAR

AUTomotive Open System ARchitecture (AUTOSAR) is a consortium of automotive companies founded in the year 2003 with an objective to standardize similar functionalities under an open, standardized software architecture for the automotive Electronic Control Units (ECUs). This consortium was founded by automotive partners like Bavarian Motor Works (BMW), Continental AG, Robert Bosch GmbH, Siemens VDO, Daimler AG (formerly Daimler-Benz, then DaimlerChrysler), and Volkswagen.

A modern-day automotive vehicle contains numerous ECUs, and the task of writing the software components for every single ECU of them right from the scratch is a tedious and time-consuming process [1]. This could be resolved by developing a standard that comprises the similar functionalities under one roof, so that they can be reused wherever and whenever necessary.

The motto behind AUTOSAR is to provide a set of specifications of the basic software modules [2] and application interfaces and to build a common infrastructure for the automotive manufacturers to comply with.

By this, the basic software modules [2] developed based on AUTOSAR standard can be implemented in vehicles and electronic components of different manufacturers resulting in the reduction of research and development expenditures, conservation of resources, mastering of the ever-evolving complexity of automotive software architecture, and promotion of reusability [3] of the software, thus saving time, effort, and resources. The AUTOSAR-compliant software architecture developed is

Y. Catherine Yamili (✉)
Robert Bosch Engineering and Business Solutions Private Limited, Coimbatore, India

M. Kathiresh
PSG College of Technology, Coimbatore, India

M. Kathiresh, R. Neelaveni (eds.), *Automotive Embedded Systems*,
EAI/Springer Innovations in Communication and Computing,
https://doi.org/10.1007/978-3-030-59897-6_3

independent of the hardware used and can be adapted to different vehicles, platforms, and other hardware [4].

Though the software modules are standardized, AUTOSAR supports extensibility. The AUTOSAR-compliant software modules can be extended in their functionalities yet with a consideration of their configuration in the automatic basic SW configuration process. The nonstandard modules can be included as the Complex Drivers. The layered architecture of AUTOSAR is discussed as follows.

Architecture

AUTOSAR follows a layered software architecture [5] where the hierarchical structure is travelled in a top-down approach. The broadest abstracted levels that the AUTOSAR architecture divides the embedded software into are three: the Application Layer, the Runtime Environment (RTE), and Basic Software (BSW).

The Application Layer

The Application Layer is the topmost layer in this three-layered AUTOSAR architecture and plays the important role in the vehicle applications. The AUTOSAR Application Layer comprises a number of application-specific software components that are used to execute tasks specified by the user. The Application Layer implementation is done using a "component" concept, and three important components to be considered in this implementation are as follows:

- The AUTOSAR application software components
- The AUTOSAR ports
- The AUTOSAR port interfaces

AUTOSAR Application Software Components

The AUTOSAR application software components are the simplest form of an application, and the end-to-end functionality comprises many such interconnected software components (SW-C). There are entities called the Runnables which are the atomic particles of these SW-Cs, given by the component. The Runnables are the procedures that hold the actual implementation of these SW-Cs and are triggered either cyclically or due to the occurrence of an event. There are various numbers of SW-Cs available within the Application Layer. One such is the Sensor/Actuator Software Component which is depicted as shown in Fig. 1.

The Sensor or the Actuator AUTOSAR SW-C is a software component that performs sensor evaluation and actuator control. As the sensors and the actuators are

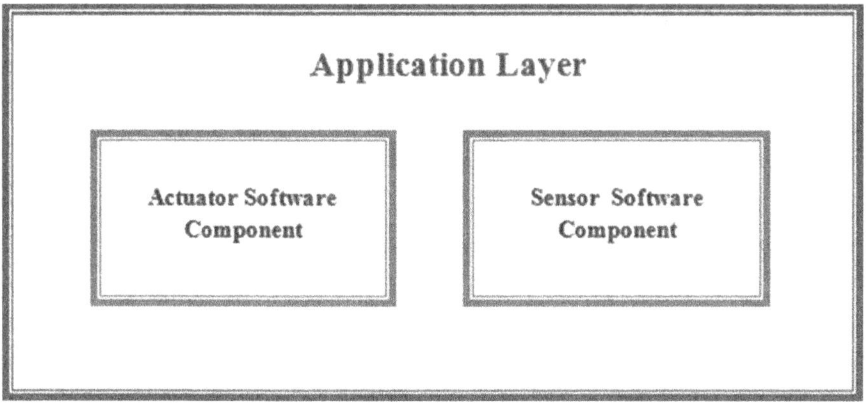

this figure will be printed in b/w

Fig. 1 Application layer SW-C

associated with the local signals, they are included in the Application Layer software components rather than being included in the Basic Software. The Sensor/Actuator AUTOSAR Software Component is placed above the RTE for integration reasons, as the SW-C can strongly interact with the raw local signals and provide an abstraction from the physical properties of the external peripherals, hardware sensors, and actuators, which are connected to the ECU hardware layout.

The AUTOSAR Ports

A port represents a communication point between the components and is mapped to a single component. The AUTOSAR SW-Cs use well-defined ports. The Virtual Function Bus (VFB) manages the inter-component communication.

The AUTOSAR Interfaces

There are standardized interfaces for the application software components. The concept of interfaces was introduced so that the communication of the data or the required services through a port of a component can be made better. The interface serves as the input to the RTE port creation. The interface is divided as follows.

- **Client-Server Interface**

  The client usually initiates the communication by requesting the server of a service. The server responds with a response message after performing the service. The AUTOSAR SW-C can either be a client or a server and is identified by the direction of the message.

- **Sender-Receiver Interface**

This interface is used where a more data-oriented informational exchange has to take place. This interface defines an asynchronous distribution of information. Here, the interface decides on what information has to be exchanged and the type of the services to be called by the client-server communication.

Real-Time Environment (RTE)

The RTE is the layer below the Application Layer and is responsible for the provision of communication services to the AUTOSAR SW-Cs and/or AUTOSAR Sensor components and the Actuator components. On providing these communication services, this layer enables the AUTOSAR Software Components to become independent of the mapping process to a particular ECU.

Basic Software (BSW) Layer

The fundamental objective of the BSW layer is that it holds the responsibility for the management of the hardware resources and the provision of common sources for the application software. It can be entitled as a standardized software module, which comprises modules that are specific to ECU and the generic AUTOSAR modules, which are all put together to aid in the functioning of the upper layer. The BSW layer is again subdivided into four layers which is shown in Fig. 2. The layers of the BSW layer are as follows:

- The Services layer
- The ECU (Electronic Control Unit) abstraction layer
- The Microcontroller Abstraction Layer (MCAL)
- The Complex Drivers

Basic Software Module Types

Driver

A driver is a software program that operates or controls the device that is being attached to the controller. The purpose of the driver is to provide a software interface to the hardware devices enabling them to gain access to the operating system and other computer programs though not being aware of the details on the hardware being used. This hardware device maybe either internal or external to the controller depending on which the drivers are classified as the internal or the external drivers.

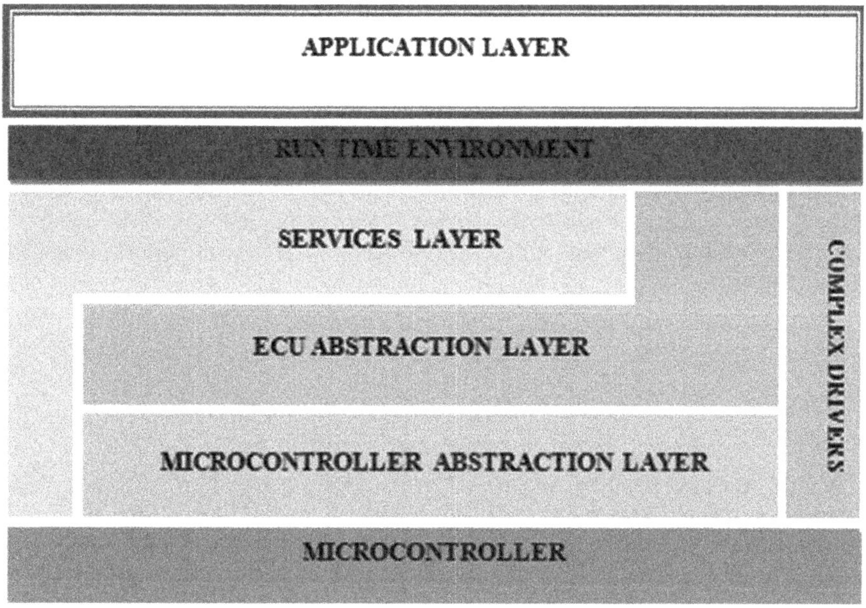

this figure will be printed in b/w

Fig. 2 Basic software layers

Internal devices are devices that are present within the microcontroller. A few examples for internal devices are internal electrically erasable programmable read-only memory (EEPROM) and internal analog-to-digital converter (ADC). Internal drivers are drivers for these internal devices and are located within the Microcontroller Abstraction Layer. Similarly, there are external devices found outside the microcontroller, located on the ECU hardware. A few examples for these external devices are external erasable programmable read-only memory (EEPROM) and external watchdog. A driver for an external device is called the external driver. The external drivers are located in the ECU Abstraction Layer, and they use the drivers present in the Microcontroller Abstraction Layer to gain access to the external devices. The components like the transceivers and watchdogs that are integrated into the System Basis Chips (SBCs) are indeed supported by AUTOSAR.

Interface

An interface acts like a boundary that is being shared by two or more components of the computing system to exchange information. The interface is used to achieve abstraction. The interface modules are designed to abstract from the modules that are placed them in the architecture. They provide a generic API, with which the access to a particular type of device can be gained. The interface modules do not tamper the data and are located in the ECU Abstraction Layer.

Handler

The functionality of a handler is incorporated in the interface or the driver modules. The handler is a specific interface module that is designed to control various types of accesses, namely, concurrent access or multiple access or even asynchronous access of a single or multiple clients to one or many drivers. In order to provide these types of access, the handler is capable of performing operations like queuing, arbitration, multiplexing, and buffering. The handler does not modify the contents of the data.

Manager

When multiple clients need specific services, the manager comes into play. The manager is used in cases where the functionality of a pure handler is not sufficient enough to abstract from multiple clients. The content of the data can be evaluated and changed or adapted accordingly by the manager. The managers are all located in the Services Layer. One among them is the Non-Volatile Random Access Memory (NVRAM) manager. The NVRAM Manager provides concurrent access to both the internal and external devices. Besides that, it also manages data checking, reliable and distributed data storage, provision of default values, and various other operations.

Libraries

Libraries comprise a set of functions for specific purposes. The AUTOSAR libraries provide other BSW modules and application software components with mathematical services. The libraries offer C functions that can be called from a source code, i.e., from BSW modules, from SW-C, from RTE, or from Complex Drivers.

Libraries:

- can be called by BSW modules (that including the RTE), SW-Cs, libraries, or integration code
- run in the context of the caller in the same protection environment
- can only call libraries
- are re-entrant
- do not have internal states
- do not require any initialization
- are synchronous

Some of the libraries which are specified within AUTOSAR are as follows:

- Fixed-point mathematical
- Floating-point mathematical

- Interpolation for fixed-point data
- Interpolation for floating-point data
- Bit handling
- E2E communication
- CRC calculation
- Extended functions

Software Layers: Overview

The Services Layer

The Services Layer is the topmost layer of the BSW layer of the AUTOSAR architecture. This is in accordance with the ease of provision of the basic software modules from the BSW layer to the Application Layer. The implementation of this layer is done in such a way that it mostly remains independent of the microcontroller and the ECU hardware. Its prime task is to provide basic services for the RTE, the layer above the BSW layer, applications, and for basic software modules. The functions of this layer are listed:

- Operating system functionality
- Vehicle network communication and management services
- Memory services (NVRAM management)
- Diagnostic services (including UDS communication, error memory, and fault treatment)
- ECU state management, mode management
- Logical and temporal program flow monitoring

The ECU Abstraction Layer

The main task of this ECU Abstraction Layer is to make the higher software layers independent of the ECU hardware. As a result, this layer acts as an interface to the drivers of the Microcontroller Abstraction Layer. Besides acting as an interface, it also contains drivers for the external devices. The peripherals and the devices irrespective of their location, i.e., within the microcontroller or external to it, and their connection to the microcontroller, i.e., whether the devices are connected to the port pins of the microcontroller or via any interface, can be accessed through an API provided by this layer.

The Microcontroller Abstraction Layer (MCAL)

The lowest layer of the BSW layer is the Microcontroller Abstraction Layer. The MCAL, just like its above layers, makes the higher layers of the BSW layer independent of the microcontroller. The MCAL comprises the internal drivers that are the software modules which enable direct access to the internal peripherals and the microcontroller. The internal structure of the MCAL is shown in Fig. 3. The drivers present in the MCAL are as follows:

- Microcontroller Drivers which are the drivers for the internal peripherals like watchdog timer and general purpose timers
- Communication Drivers, the drivers for the ECU on-board and Vehicle Communication
- Memory Drivers for On-Chip Memory devices like the internal EEPROM and external memory devices like the external flash
- Input/Output Drivers for the Analog and Digital I/O applications like ADC and PWM
- Crypto Drivers for crypto devices
- Wireless Communication Drivers for the wireless network systems in case of off-board communication

Complex Drivers

The Complex Driver is the module that stretches right from the RTE to the microcontroller with the purpose of exhibiting a nonstandardized, i.e., the functionalities that are not AUTOSAR-specific, and special-purpose functionality within the basic software stack. These come up with high timing constraints. The AUTOSAR-layered architecture works in a way that the access to the hardware devices from the upper layers is abstracted. This rendered restriction to access the critical resources or the non-AUTOSAR-compliant SW-Cs. Hence to overcome this, the Complex Drivers were designed. External devices like complex sensors could be integrated and evaluated with direct access to the microcontroller by making use of these Complex Drivers.

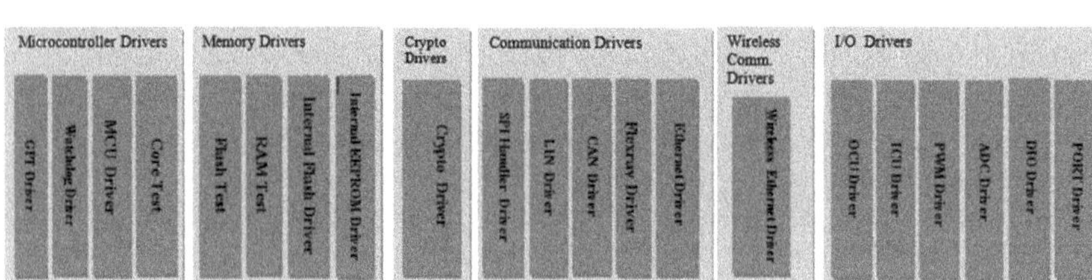

Fig. 3 Internal structure of MCAL

Software Layers: Functional Groups

The Basic Software provides a number of various services. Hence every layer of the BSW layer is again divided into further functional subgroups, each group accounting for services like Input/Output (I/O), Memory, Crypto, Communication, Off-Board Communication, and System Services which are dealt with in detail.

I/O Hardware Abstraction and Services

The inputs and the outputs that are directly connected to the ECU and that which do not have exclusive drivers dedicated to them are all handled by the I/O Hardware Abstraction [6], which is a group of modules designed for this purpose shown in Fig. 4. These inputs and outputs are either directly mapped to the ports of the microcontroller or to an on-board peripheral. The communication between the peripherals and the microcontroller except those external devices that are being handled by the Complex Drivers is abstracted by the I/O Hardware Abstraction. All these Input and Output devices can be accessed through the I/O signal interface; hence the inputs and the outputs are observed as the electrical signals. It performs static abstraction and inversion of values according to their physical representation at the inputs/outputs of the ECU hardware [7]. The ECU hardware and its properties are being abstracted such that they remain hidden to the upper software layers so that the burden of the dependency on these layers is lifted off from the higher layers.

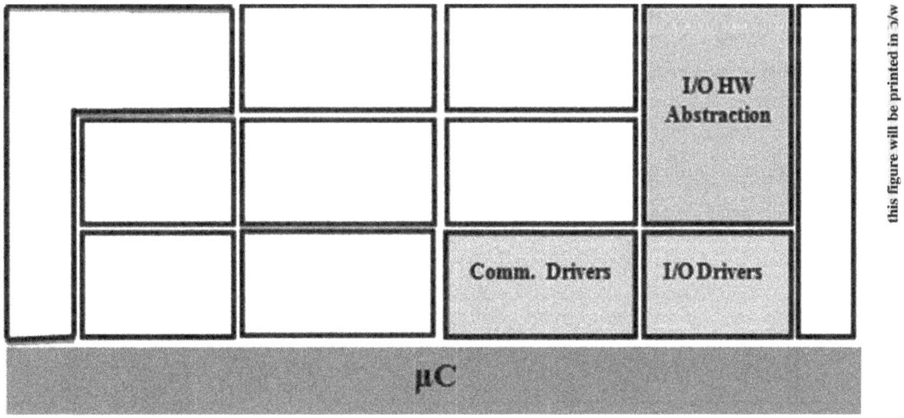

Fig. 4 I/O hardware abstraction

Memory Hardware Abstraction and Services

The Memory Hardware Abstraction module, shown in Fig. 5, abstracts from the addressing scheme of the underlying memory hardware drivers and provides a uniform addressing scheme. It thus supports a virtually unlimited number of read and write cycles since it supports schemes like reconfiguration of memory. By doing this, the upper layer can remain unchanged even if the layers lying are changed. The dependency on the lower layers is lifted off from the above layers. By this way, equal mechanisms are being employed to access both the internal and the external memory devices and any other type of memory hardware.

The Memory Services comprise one module known as the NVRAM (Non-Volatile Random Access Memory) Manager which is shown in Fig. 6. As the name goes by, the prime task of the memory services is that it takes up the responsibility of managing the non-volatile data. The application is provided with the non-volatile data in the uniform way by this manager. The non-volatile data and the mechanisms acted upon it like saving, loading, checksum protection and verification, and reliable storage are being taken care of by this NVRAM Manager

Crypto Hardware Abstraction and Services

The Crypto Driver is responsible for the cryptographic implementations. It also supports the key storage, configuration, and management for the cryptographic services [8]. The Crypto services comprise one module known as the Crypto Service Manager (CSM) [8]. It controls the access of single or multiple clients to one or more cryptographic services, be it synchronous or asynchronous services. The CSM uses prioritized queues to store the tasks to which the dedicated Crypto Driver could pay attention. Figure 7a, b shows the Crypto Hardware Abstraction and Services,

this figure will be printed in b/w

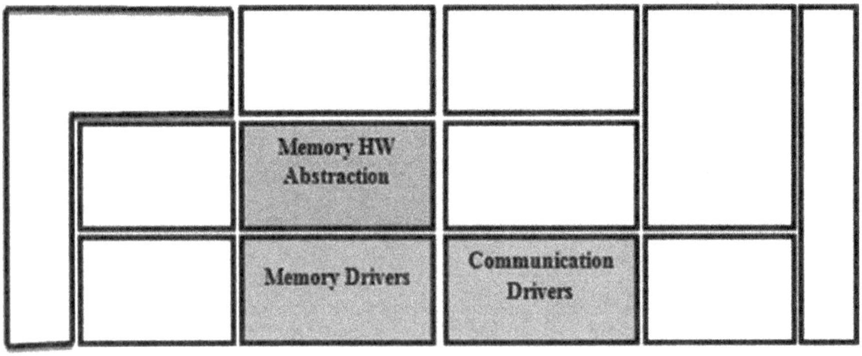

Fig. 5 Memory Hardware Abstraction

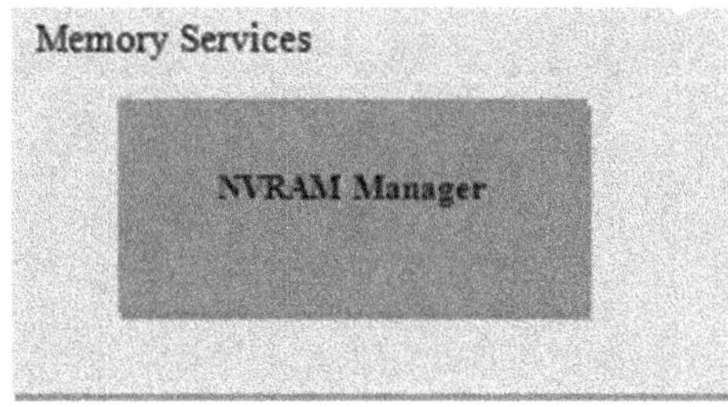

Fig. 6 NVRAM Manager

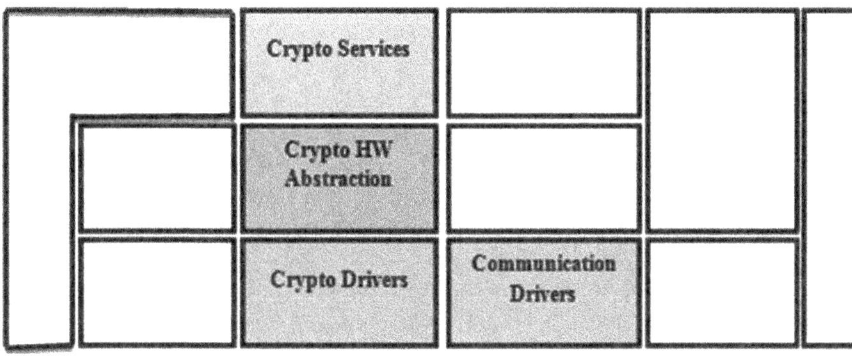

Fig. 7 (**a**) Crypto Hardware Abstraction and Crypto Services. (**b**) Crypto Hardware Abstraction

and Fig. 8 shows the Crypto Services and the Crypto Service Manager. The functionality of the CSM are listed as follows:

- Hash calculation
- Generation and verification of message authentication codes
- Generation and verification of digital signature
- Encryption and decryption using symmetrical or asymmetrical algorithms
- Random number generation
- Secure counter
- Key management operations, e.g., key setting and generation

The services of the CSM are quite generic, and besides that CSM also allows different applications to use the same service with different cryptographic algorithms. The service client need not worry over the cryptographic routines, if they are implemented in the software or the hardware or which Crypto Driver is being requested to take care of, because all of these are encapsulated by the CSM as it provides an abstraction layer [9] to all of these features.

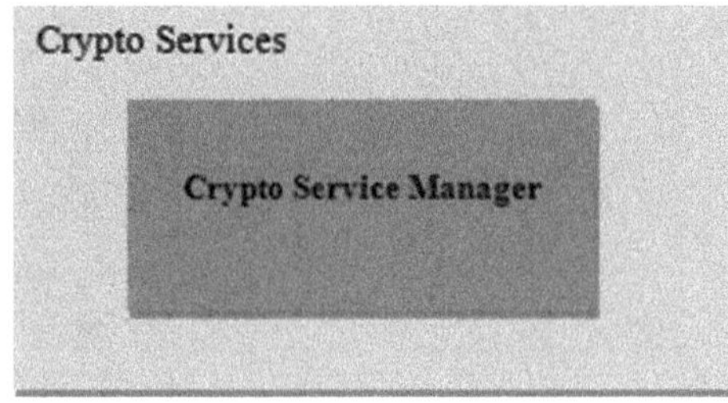

Fig. 8 Crypto Service Manager

Communication Services

The Communication Hardware Abstraction is a collection of modules that performs abstraction from the location of communication controllers and the ECU hardware. Communication systems like CAN, LIN, and FlexRay require a specific Communication Hardware Abstraction. The Communication Drivers are accessed through bus-specific interfaces. A bus channel can be accessed via equal mechanisms irrespective of its location, i.e., if it is present on-chip/on-board.

The Communication Services, a group of modules, are designed to support the vehicle network communication [10]. The Communication Hardware Abstraction is through which these Communication Services gain access to the Communication Drivers. These Communication Services provide the Vehicle network with a uniform interface for communication, for diagnostic communication. They also provide uniform services for the network management. Communication Services supporting CAN, LIN, and FlexRay are discussed as follows.

Communication Stack: CAN (Controller Area Network)

The CAN Communication Services, a group of modules, are designed to take care of the vehicle network communication which uses CAN as the communication system. The prime task of the CAN Communication Services is to provide a uniform interface to the CAN network. In addition to this, the CAN Communication Stack also supports Classic CAN communication (CAN 2.0) and CAN FD communication.

There is an optional extension of the plain CAN interface and the CAN Driver module for the vehicle network communication known as the TTCAN (Time-Triggered Controller Area Network) Communication Services with the communication system TTCAN. This, alike the CAN communication services, provide a uniform interface to the TTCAN network. The TTCAN network acts like a superset to CAN, which means that a CAN stack that supports TTCAN can serve both the CAN and the TTCAN bus. Hence the CAN interface along with TTCAN can serve both the CAN Driver and CAN Driver TTCAN. This also makes the properties of the CAN stack true for CAN with TTCAN functionality as well. The CAN Communication Stack is microcontroller and ECU hardware independent but partly dependent on CAN. The Generic Network Management (NM), AUTOSAR COM, and Diagnostic Communication Manager remain the same for all the communication systems of the vehicle network systems. The Generic NM contains a dispatcher and sometimes may also include NM coordinator to synchronize multiple networks. The CAN NM remains exclusive to the CAN networks, and the CAN State Manager takes care of the start-up and shutdown features which are dependent on the communication system. The CAN Communication Stack is shown in Fig. 9.

Communication Stack: LIN (Local Interconnect Network)

- **LIN Master**

The LIN Communication Services, a group of modules, are designed to take care of the vehicle network communication which uses LIN as the communication system. The prime task of the LIN Communication Services is to provide a uniform interface to the LIN network. The following are the properties of the LIN Communication Services:

A LIN 2.1-compliant communication stack with

- Schedule table manager which takes care of transmitting LIN frames and handling requests to switch to the other schedule tables
- Transport Protocol that is used for the diagnostic purpose
- A Sleep-Wake Up Interface

A LIN Driver that is used for

- Implementing the LIN protocol
- Supporting both the simple UART and the complex frame-based LIN hardware
- **LIN Slave**

The LIN slaves are the intelligent actuators that are used for any kind of control mechanisms and slaves that are perceived as black boxes. In certain scenarios LIN slaves are also used as ECUs. But due to the limited hardware capabilities, they are not forced to comply to the AUTOSAR SW architecture.

They usually have limited memory resources. All of the abovementioned reasons are sufficient enough to tell that it is not advisable to shift AUTOSAR SW-C into them. The LIN interface holds control over the Sleep/WakeUp API and provides privileges to the slaves to keep the bus awake in a decentralized approach. The LIN NM remains exclusive to the LIN networks.

The LIN State Manager besides handling the start-up and shutdown features which are dependent on the communication system also holds control over the communication mode requests from the Communication Manager. Figure 10 shows the Communication Stack of LIN.

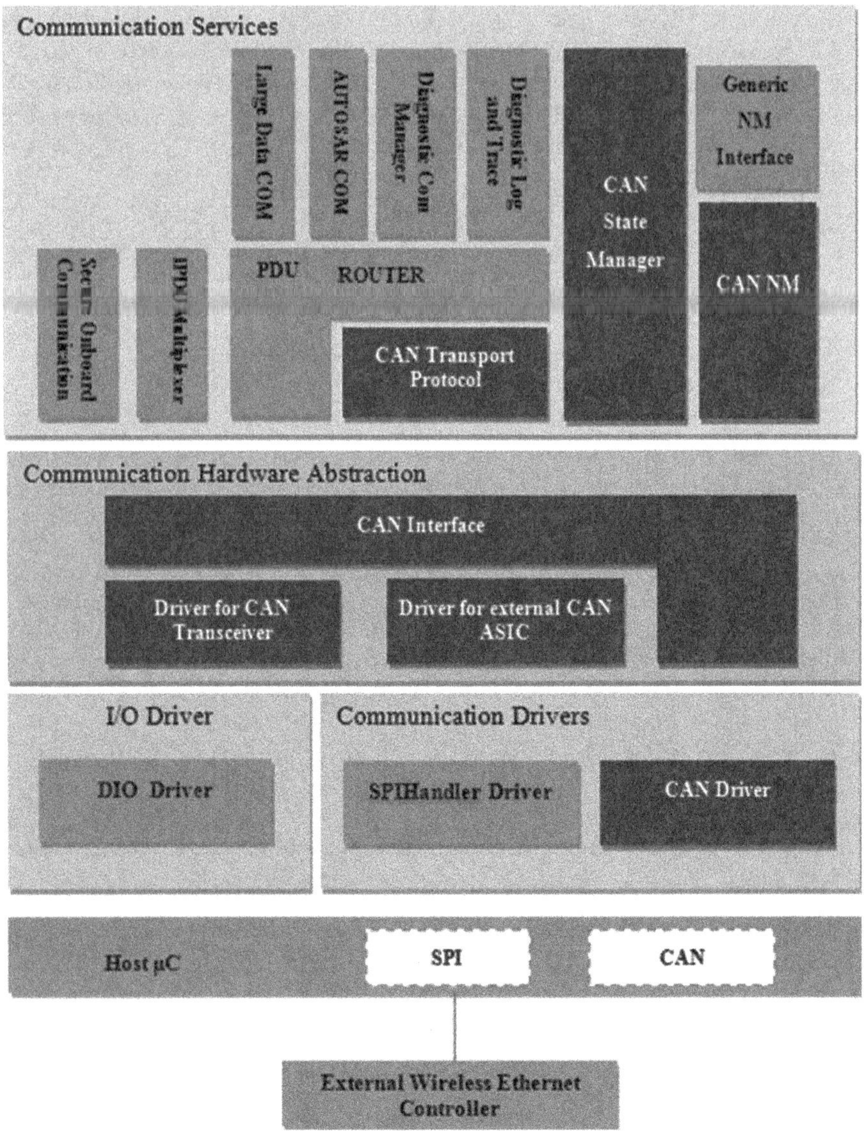

Fig. 9 Communication Stack: CAN

Communication Stack: FlexRay

The Communication Services of the FlexRay component are a group of modules that are designed to take care of the vehicle network communication which uses FlexRay as the communication system. The prime task of the FlexRay Communication Services is to provide a uniform interface to the FlexRay network.

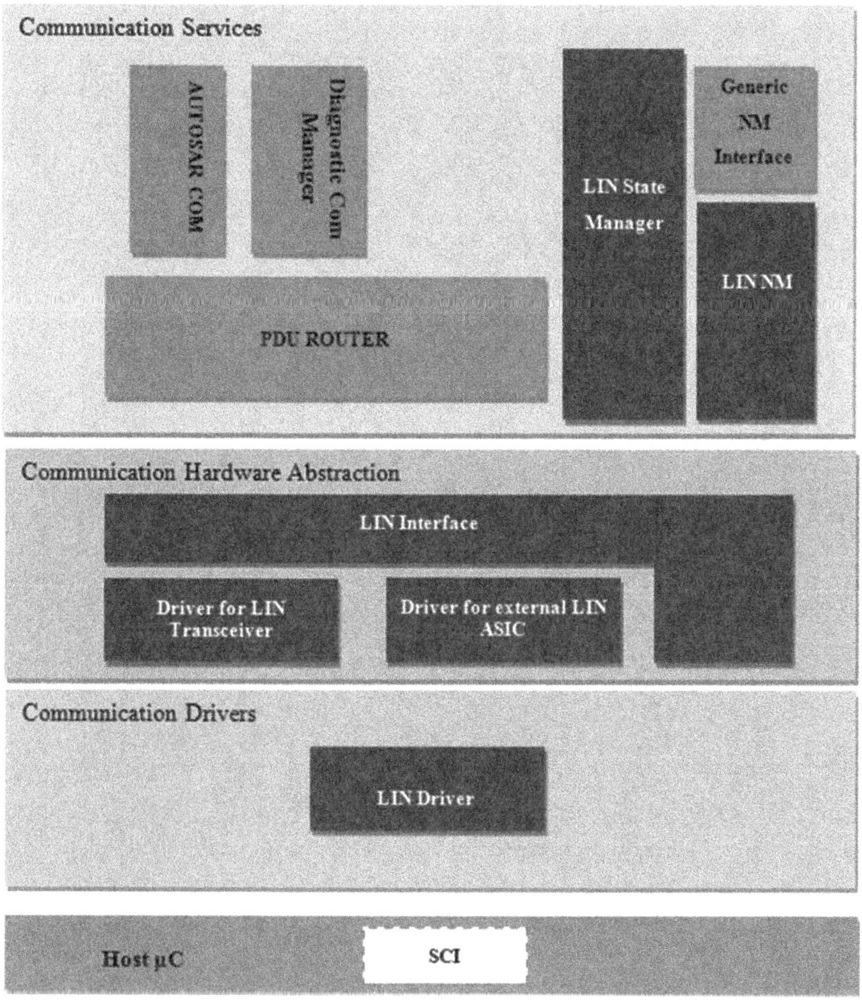

this figure will be printed in b/w

Fig. 10 Communication Stack: LIN

The FlexRay Communication Stack is microcontroller and ECU hardware independent but partly dependent on FlexRay.

The FlexRay NM remains exclusive to the FlexRay networks. The FlexRay State Manager controls the different options of COM to send the Protocol Data Units (PDUs) and monitor signal timeouts. Figure 11 shows the Communication Stack of FlexRay.

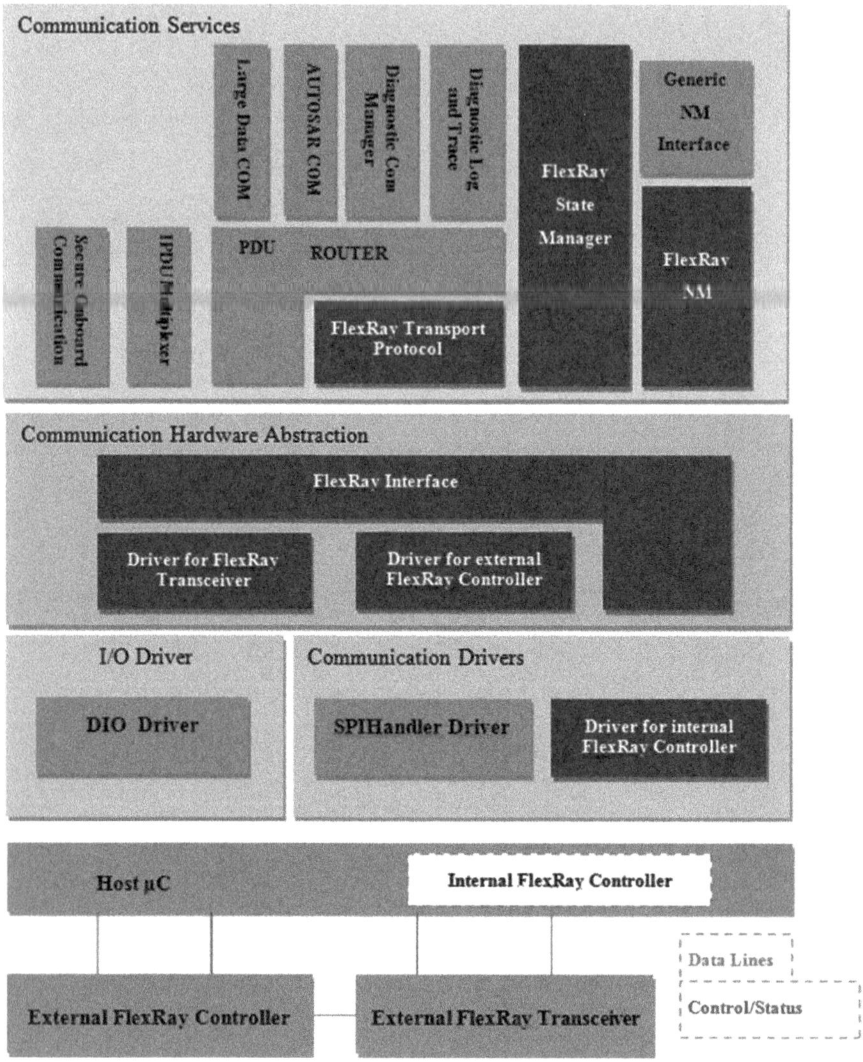

Fig. 11 Communication Stack: FlexRay

Communication Stack: TCP/IP

The Communication Services of TCP/IP are a collection of modules that are designed to take care of the vehicle network communication which uses TCP/IP as the communication system. The TCP/IP Communication Services provide a uniform interface to the TCP/IP network. The protocols that the TCP/P module implement are TCP, UDP, IPv4, IPv6, CMP, ARP, and DHCP. Besides this, it also supports a dynamic, socket-based communication via Ethernet. The Socket Adapter (SoAd) module serves as the upper layer module of the TCP/IP module.

Off-Board Communication Services

A provision for the Vehicle-to-X Communication, where the X can be replaced with another vehicle or grid, etc. through an ad hoc wireless network, is being achieved by this Off-Board Communication Services.

These are a group of modules designed for the above purpose. The Off-Board Communication Services offer certain facilities like the functionality which receives and transmits the standardized V2X messages and builds the vehicle-specific software components' interface. It implements the Basic Transport Protocol in Layer 4 of the OSI model, i.e., the Transport Layer, and stations the Geo-Networking in Layer 3, the Network layer, where the addressing is based on the geographic areas and the respective Ethernet frames have their own EtherType. The V2X Management manages the cross-layer functionality which includes the dynamic congestion control, security, position, and time. The Off-Board Communication Services hide the properties of the messages and the protocol and its related information from the application, besides providing a uniform interface to the wireless Ethernet network. Figure 12 shows the Off-Board Communication Services.

On-Board Device Abstraction

This On-Board Device Abstraction comprises the drivers for the ECU on-board devices apart from the sensors and the actuators like the internal or the external watchdogs. Those drivers access the ECU on-board devices through the microcontroller abstraction layer. Figure 13 shows the On-Board Device Abstraction.

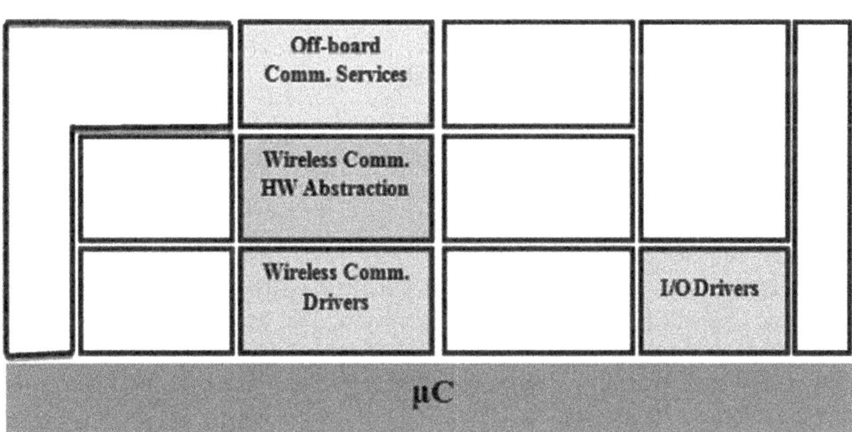

this figure will be printed in b/w

Fig. 12 Off-Board Communication Services

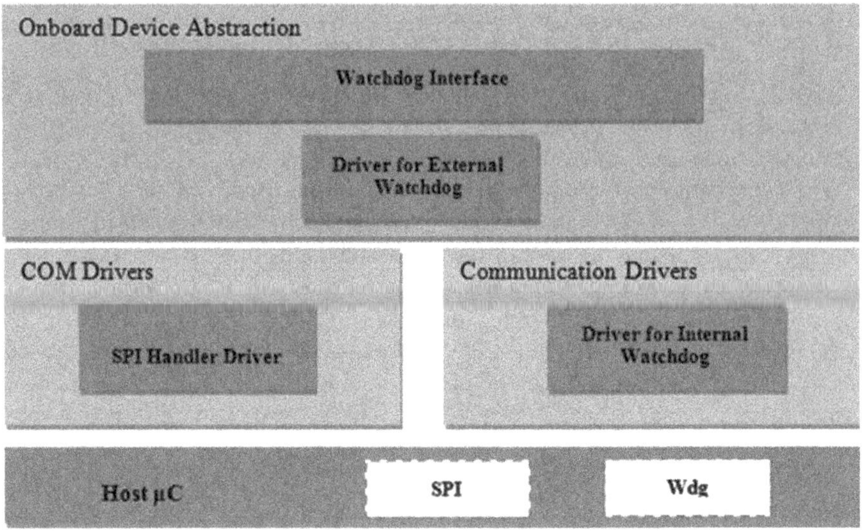

Fig. 13 On-Board Device Abstraction

System Services

There are a set of modules and functions that are designed in such a way that they can be used by the modules of all the layers. This group of modules is called as the System Services. Some of these services are microcontroller dependent, like OS services and support special microcontroller capabilities like Time Service, whereas some of them are partly ECU hardware and application dependent services like ECU State Manager and some of them are hardware and µC independent services. Figure 14 shows the System Services.

Error Handling, Reporting, and Diagnostics

Every aspect of error handling in AUTOSAR has a dedicated module associated to it. Consider the case of the Diagnostic Event Manager. It processes and stores the diagnostic events and the associated Freeze Frame data. The Diagnostic Log and Trace module takes care of the logging and tracing of the applications. The user-defined log messages are collected and converted into a standardized format by this Diagnostic Log and Trace module. The Default Error Tracer is being reported with the detected development errors in the Basic Software. A common API for diagnostic services is provided by the Diagnostic Communication Manager. The error handling modules are seen in Fig. 15.

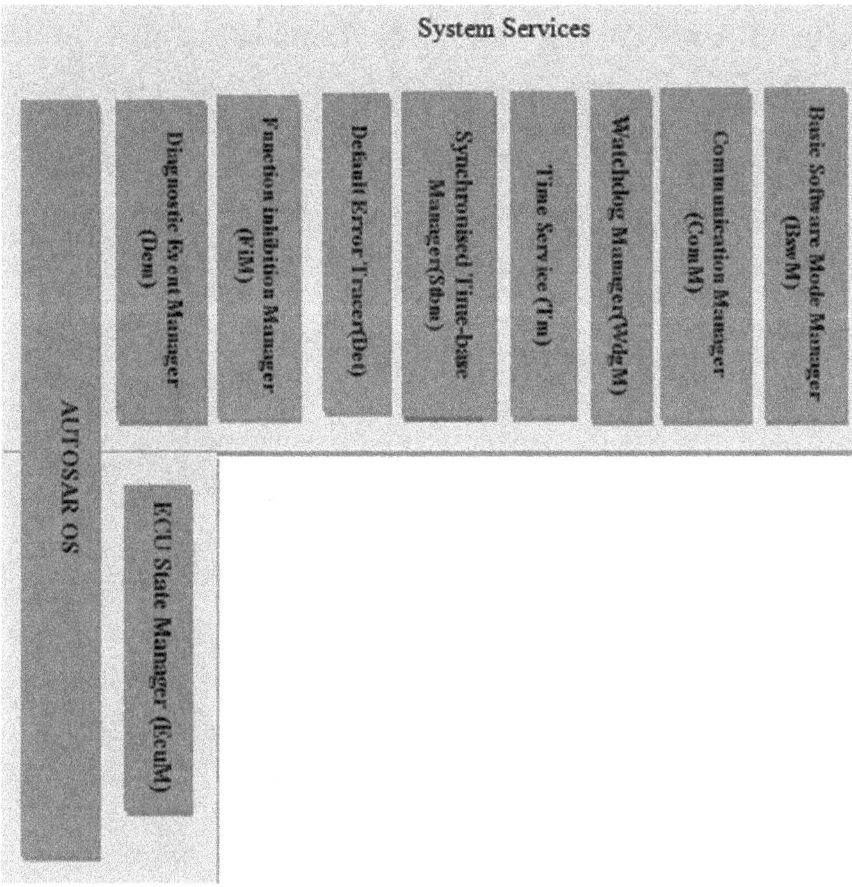

this figure will be printed in b/w

Fig. 14 System Services

Communication Between the Software Components

The Virtual Function Bus (VFB) is an abstract component that abstracts the application from the architecture and manages the communication between the AUTOSAR Software Components. It is represented by the RTE. For every ECU present in the AUTOSAR system, the VFB is generated exclusively. The VFB gets to choose the ports for communication, and the communication interfaces of the SW-Cs should be mapped to the ports for the communication to take place. Besides that, the VFB also handles communication between and within the ECUs, and Fig. 16 shows both Inter- and Intra-ECU Communication.

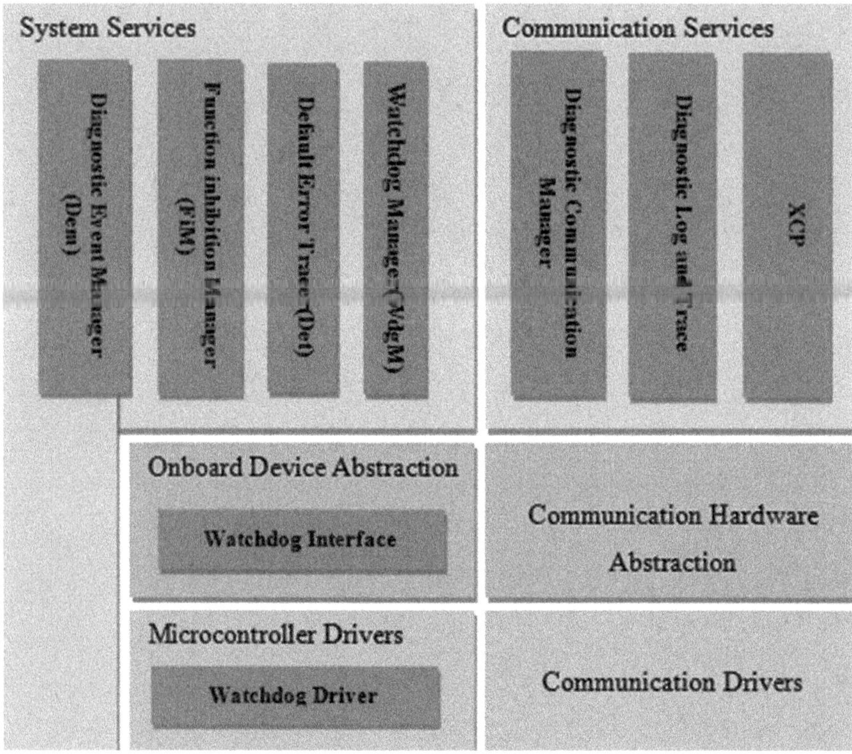

Fig. 15 Error handling modules

Inter-ECU Communication

Inter-ECU Communication takes place through the Real-Time Environment and proceeds further into the Basic Software module where the BSW manages the functions, namely, the diagnostics, and takes care of the memory services, communication services, etc. In Fig. 16, the communication between the ECU I and the ECU II stands as an example of Inter-ECU Communication.

Intra-ECU Communication

The communication between the software components is termed as Intra-ECU Communication, and all these components are mapped to the same ECU. This takes place entirely through the RTE. The SW-C B and C in Fig. 16 communicate using the Intra-ECU method through the RTE.

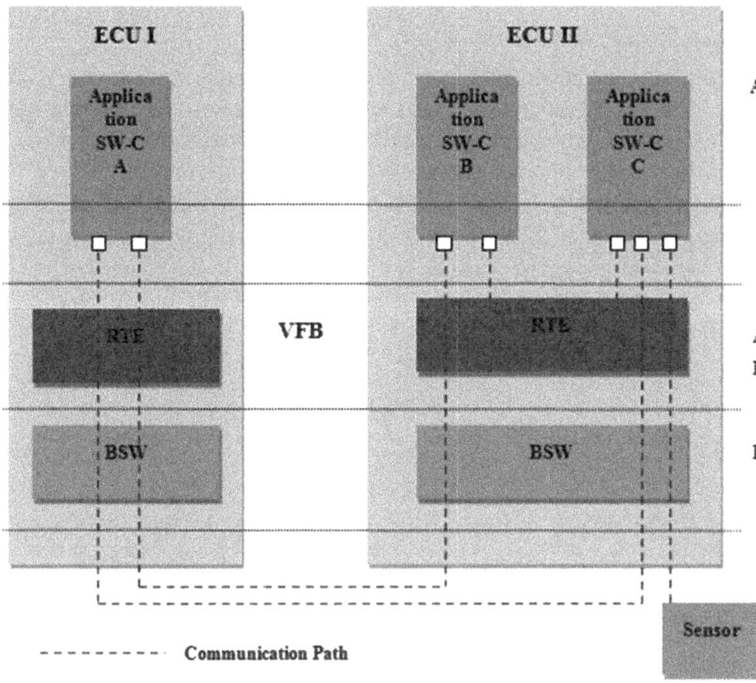

Fig. 16 Intra- and inter-ECU communication

this figure will be printed in b/w

Overview of MISRA

On the onset of the 1990s, the field of electronics started its extension to a greater percentage into the field of the automobiles, and with this, the software used gained importance in the progressive states. As and when the software used evolved, the software reliability posed higher impacts to the automobiles' performance. As a result, the way of software coding is taken into consideration to minimize the reli ability issues as they are a big deal when it comes to handling critical resources. MISRA, the Motor Industry Software Reliability Association, is a consortium formed from the representatives of the companies of the automotive industry [11]. It was set up in the year 1998 initially by the UK Government to set up guidelines on the development of embedded software.

The most widely used programming languages in the development of the embedded software are C and C++, as C besides providing a diverse range of libraries and resources also is supported by a wide range of tools. Though a C program conforms and compiles to the ISO language standard, the code may still exhibit unpredictable behavior, which certainly cannot be tolerated in safety-critical systems. But with the help of MISRA coding guidelines, the code when made to conform to it may substantially reduce uncertain behavior, so that the dangerous aftereffects posed by the same can be contained.

The coding standards define a set of consistent coding practices, a safer subset of the actual programming language, eliminating the practice of using coding constructs that might result in hazardous situations or in undefined behavior. One such instance is an arithmetic operation that causes an integer overflow while computing the result. The motto of the coding rules prescribed by these standards does not stop with just identification but prevention as well. The coding rules prevent the developer from developing code which might produce unpredictable results by restricting the use of the language. The MISRA coding rules are widely accepted as the benchmarks used in the development of the software deployed in the safety-critical systems [11]. Besides being concise and readable, they also focus on the essential issues making them ubiquitous.

MISRA C: 1998

The first edition of the MISRA coding guidelines "Guidelines for the Use of the C language in Vehicle Based Software," was published by MISRA in the year 1998. This edition consisted of 127 coding rules [12]. The foundation was laid by two MISRA members Ford and Rover.

MISRA C: 2008

The wide acceptance and adoption of MISRA C: 1998 motivated the founders to improvise the coding standard; as a result of which, the areas that needed further improvisation and clarification were identified and worked upon. The revised version of the same was published in the year 2004. Though the structure was modified and additional rules were added, the flavor of the original version was preserved. The title was modified to "critical systems" from just "vehicle based software."

MISRA C++: 2008

Though the programming language C was used predominantly in the safety-critical systems, there arose an increase in the usage of C++ language. Hence a new committee was established and MISRA C++ was formed in the year 2008.

MISRA C: 2012

MISRA C: 2012, with further improvisation that has been garnered from 14 years of experience drawn from organizations across the world, in the guidelines, was published in the April 2013 [13]. It also extends its support to the C99 version of the C language besides the addition of improvements that would result in the reduction of cost and complexity of compliance while still adhering to the concept of consistency and reliability of C in the safety-critical systems.

MISRA: Safe and Secure

Perceptions prevail stating that MISRA C is related only as an aid of safety measures and not of security measures as a MSIRA C defines a language subset. Safety-related standards need the deployment of a language subset, and as MISRA C defines a safer subset of the actual programming language, thus being highly associated with integrity and reliability requirements. But, to prove that MISRA is also equally associated with security-related applications, MISRA C carried out a comparison in response to the publication of ISO/IEC 179161:2013, the C language security guidelines which was published by the C standard committee. The comparison result was published as Addendum 2 to MISRA C: 2012, and this comparison threw light on the areas where MISRA C could be enhanced, resulting in MISRA C: 2012 Amendment 1. This amendment extends MISRA C: 2012 with the addition of new guidelines aimed at the improvement of the security concerns that were inferred from the comparison with the ISO C Security Guidelines which addressed the security vulnerabilities.

MISRA Compliance and Deviation Permits

MISRA C published MISRA Compliance: 2016 in April 2016. This document was designed on the basis that circumstances may arise where adhering to these guidelines become quite impracticable. In cases like these, the compliance document comes to aid. This document defines a framework for not only achieving compliance with the MISRA C guidelines but also has a set of guidelines on a structured and robust process for the use of deviations. It provides clear guidelines on the use of the deviations. This document permits approved violations to match the need of the project.

Rules

The rules in the MISRA document [11, 14] are categorized under different topics which are explained. Sometimes an overlap may arise while categorizing, where the rule is often found to come under more than one topic. In these circumstances, the rule is placed under the most relevant topic. The classification of these rules is that they either come under "required" or "advisory."

Required Rules

The rules that are classified under this division are mandatory rules that have to be followed by the programmer. The code that is said to be developed by the programmer has to conform to every rule classified under this division.

Advisory Rules

These rules are to be followed by the programmer normally. But they are not posed with the mandatory status as that of the required rules. Though they do not have the mandatory status marked on them, it does not mean that they can be ignored. It simply means that they can be followed till they become impractical or unreasonable and deviations can be raised if considered appropriate.

Presentation of Rules

The rules in the MISRA C document are presented in the following format.

```
Rule <number> (<category>): <requirement text>
```

The fields in the rule format are described as follows:

<number>	Every rule is associated with a unique number that is followed sequentially throughout the document irrespective of the section they come under
<category>	This field depicts if the respective rule comes under the classification of either "advisory" or "required."
<requirement text>	The rule statement

The rules are accompanied with supportive statements, which talk about the application of the rules and the underlying issues associated with the rules. **There** are about 143 rules, inclusive of both the required and advisory categories, and few of these rules under different categories [15] are explained with suitable examples in the forthcoming sections.

Rules

Environment

Provisions should be provided to look out for the occurrence of run-time errors. A special attention is required, because C by itself is not equipped with this handling of dynamic errors, which is a very important aspect in the safety-critical systems.

Rule D.4.1 (required): Run-time failures shall be minimized

Hence, there arises a need for cautious monitoring of the occurrence of the run-time errors by the developer, by introducing checks to the code, wherever the developer finds the possibility of the occurrence of run-time errors, keeping in mind the performance and the code-size constraints. The following are some areas which can be taken into considerations where the run-time errors might occur.

- **Arithmetic Errors**

These occur as the resultant of the evaluation of expressions giving rise to divide by zero error, overflow, or underflow scenarios. A supporting example 10.1 showing a divide by zero condition is shown.

Noncompliant Code Example 10.1

```
{
num1 = 10;
for(i=0;i<10;i++)
{
result = num1/i; // Noncompliant
}
```

In the example, the literal *i is* initially initialized to zero; hence on dividing *num1* by *i,* the zero division error is encountered with.

- **Pointer Arithmetic**

This deals with the calculation of addresses where care must be taken to ensure that the addresses computed should be reasonable and point to places meaningful.

- **Left Shifts**

Care must be taken while performing left shifts, as the Most Significant Bit is lost resulting in the overflow.

- **Array Bound Errors**

The array indices while being used for indexing the array should be ensured that they are within the bounds specified.

Character Sets

Rule R.1.3 (required): There shall be no occurrence of undefined or critical unspecified behavior

The ISO Standard defines about 91 characters as a minimum source character set that is being supported by all the compilers. The rule says that only these characters are to be used even if the compiler in use supports an even larger set. The defined escape sequences are \n, \t, \v, \b, \r, \f, \a, \\, ?, \', \", \<Octal Number>, and \x<Hexadecimal Number>. The usage of escape sequences other than these would result in unpredictable behavior. A noncompliant code example 10.2 is shown as follows.

Noncompliant Code Example 10.2

```
const char_t a[ 2 ] = "\k"; // Noncompliant
const char_t b[ 2 ] = "\b"; // Compliant
```

Comments

Rule D.4.1 (required): The character sequences /* and // shall not be used within a comment

The nesting of comments is not something that is supported in C. The comment starts on detecting /* and ends on the first occurrence of */ irrespective of any nesting of comments being attempted. A noncompliant code example supporting the rule statement is shown as follows.

Noncompliant Code Example 10.3

```
/* This is a multi-line comment statement.
/* Nesting of comments leads to unexpected behavior*/
The actual comment ends here */
```

The example shows a multi-statement comment section which is begun with /*, but in the very next line, there is a comment line ending with */, whereas the actual comment ends in the next line. *Now* this might confuse the developer as which the actual ending of the comment might be, as the final line gets uncommented, due to the early detection of the */. Hence nesting of the comments is not used.

Constants

Rule R.7.1 (required): Octal constants should not be used

Any integer constant that begins with "0" (zero) is termed as an octal constant. The usage of the octal constants is denied, because this may cause ambiguous scenarios. Example 10.4 illustrates this rule. Consider a user who wants to assign fixed length integer constants to variables. The array initialization is done as follows.

Compliant Code Example 10.4

```
data [0] = 500; /* set to decimal 500*/
data [1] = 071; /* set to decimal 57*/
data [2] = 520; /* set to decimal 520*/
```

In the code snippet, the user assigns the decimal value "071" to data [1], thinking that it would be taken in as an integer constant, but the **compiler** recognizes it as an octal value as it has begun with a "0" and hence assigns the decimal value of the octal constant "071" which is "57" to data [1]. Therefore it is better not to use octal constants, yet zero is an exception as it holds the same value even in the octal representation.

Declarations and Definitions

Rule R.8.1 (required): Types shall be explicitly specified

Implicit typing of variables and functions though supported by some C compilers shall not be used according to the rule, as it might lead to confusions. A supporting example 10.5 which shows a noncompliant form of the rule along with its compliant solution in the comments is shown as follows.

Noncompliant Code Example 10.5

```
extern x; // Compliant solution - extern int16_t x;
const x; // Compliant solution - const int16_t x;
static fun(void); // Compliant solution - static int16_t fun(void);
```

Initialization

Rule R.9.1 (required): The value of an object with automatic storage duration shall not be read before it has been set

Normally when the integer variable is declared without any storage specifier, the default specifier it takes is, the *auto* class. These integer variables shall not be used before getting initialized, to avoid any unexpected behavior due to garbage values. In example 10.6 the variable data returned by the function would hold garbage value, due to the conditional construct, as it is not initialized before being used.

Noncompliant Code Example 10.6

```
{
int data, value;
value=0;
if (value==1)
{
data = 45;
}
return data;
}
```

The intention of this rule is that all the variables have to be initialized before being used. The initialization need not necessarily be done at the time of the declaration but at some other part of the code but before the variable is put into use.

Operators

Rule R.10.1 (required): Operands shall not be of an inappropriate essential type

There are rules that are to be followed on using the operators over the operands. One such is the rule that talks about the usage of the Bitwise operator over its operands. The built-in Bitwise operators (~, >>, >>=, &, &=, ^, ^=, |, and |=), if used on the signed integer constants, yield implementation-dependent results. The most significant of them is the left shift which, if performed on the signed integers, may cause the signed bit, which is the Most Significant Bit, to be lost in the process, which leads to erroneous results. Hence, the bitwise operators are not used against the signed integers. A supporting illustration 10.7 is shown as follows.

Noncompliant Code Example 10.7

```
if ((uint16_value & int16_data) == 0x246U)
if (~int16_data == 0x246U)
```

The example 10.7 shows a bitwise operation being performed between an unsigned and a signed integer and another bitwise operation being performed on a signed integer. Both of these operations might yield results far from the unexpected one.

Conversions and Pointers

Rule R.11.3 (required): A cast shall not be performed between a pointer to object type and a pointer to a different object type

Rule R.11.3 (required): A conversion should not be performed between a pointer to object and an integer type

These rules are extracted from a set of rules that talk about conversions from one type to another. These rules in particular talk about things that have to be kept in mind while dealing with pointers and conversions. Pointer types shall neither be made to cast into other types nor other types be cast into pointers. It is because if an address value being held by a pointer variable is assigned to an integer variable, it not be able to hold the complete value of the address due to its size. In this case, there occurs data loss, and further computations involving these variables will not yield the expected results. An example 10.8 supporting this rule is shown.

Noncompliant Code Example 10.8

```
int* data;
int addr_value = (int) &data;
```

Example 10.8 shows a pointer data variable being type cast into an integer and its value being assigned to an integer value. Data loss may occur if the *addr_value* is not large enough to accommodate the address value being assigned to it.

Expressions

Rule R.12.1 (advisory): The precedence of operators within expressions should be made explicit

This rule is from the set of rules that are designed to be followed for evaluating or forming an expression for evaluation. Precedence is normally followed for the evaluation of an expression to arrive at the intended result. But the rules of precedence can be confusing at times. Hence in order to avoid this confusion which may cause unpredictable results later, it is advisable to have a limited dependency on the rules of precedence. The expression can be evaluated according to the precedence by making use of parentheses, in case of complex statements. However, there should not be overusing of the parentheses. A noncompliant code example 10.9 is shown.

Noncompliant Code Example 10.9

```
x = a == b ? a : a - b; // Noncompliant; Compliant Usage -> x = (
a == b ) ? a : ( a - b );
x = a + b - c + d; // Noncompliant; Compliant Usage -> x = ( a + b
) - ( c + d );
x = a * 3 + c + d; // Noncompliant; Compliant Usage -> x = ( a * 3
) + c + d;
```

Control Flow

Rule R.16.3 (required): An unconditional break statement shall terminate every switch clause
Rule R.16.4 (required): Every switch statement shall have a default label
Rule R.16.5 (required): A default label shall appear as either the first or the last switch label of a switch statement
Rule R.16.6 (required): Every switch statement shall have at least two switch clauses
Rule R.16.7 (required): A switch expression shall not have essentially Boolean type

These are few rules to be followed while writing a switch statement. The *switch cases* inclusive of the *default case* mark their end when an unconditional *break* statement is introduced. The usage of the *break* statement is permitted only in the case of a *switch* statement, and an exception exists in case of an empty *case* clause. The *switch* statement should have a *default* label which can be present either at the end or at the top under a *switch* statement. The *switch* expression is prohibited in using a Boolean type, as the result of a Boolean computation is either true (1) or false (0), in which case a simple *if…else* construct would suffice as there would be only two choices to decide upon. Hence *switch* statements are expected to have at least two cases to prove their usage. These are illustrated in example 10.10.

Compliant Code Example 10.10

```
switch (value) /* value is not of type Boolean*/
{
case 0:
statement 1;
break; /* break statement is required after the case statements*/
case 1: /* empty case clause, no break is required*/
case 2;
statement 1;
break;
default:
err_flag_count = 1;
break;
}
```

Functions

Rule R.17.2 (required): Functions shall not call themselves, either directly or indirectly.

Recursive functions are not considered to be used in case of safety-critical systems. This is because the usage of the recursive functions consumes the stack space and causes the danger of depleting the same. The recursive function requires high control, else it would not be possible to determine the worst-case stack usage. A supporting noncompliant code example 10.11 is shown.

Noncompliant Code Example 10.11

```
void func()
{
printf("This is a recursive function");
func(); //Noncompliant

  }
```

Arrays

Rule R.9.5 (required): Where designated initializers are used to initialize an array object, the size of the array shall be specified explicitly

The declaration of an array without specifying the size of the array is possible. But it is not allowed according to the rules, as that might be unclear. An illustration on this is shown.

Noncompliant Code Example 10.12

```
int arr1 [ ];
int arr2 [ ] = { [0] = 1, [12] = 36, [4] = 93 };
int pirate [ ] = { 2, 4, 8, 42, 501, 90210, 7, 1776 }
```

In example 10.12, all the lines of array declaration are noncompliant as the size of the array is not specified explicitly. In the second line of declaration, the highest size defines the size implicitly, and in the third line, the total number of items defines the size, but these implicit declarations of size might be unclear, hence not used according to the rule.

Structures and Unions

Rule R.19.2 (advisory): The union keyword should not be used.

The usage of union in order to access an object in different ways may cause misinterpretation of the data. Hence this rule was designed to prevent the usage of unions in any circumstances. An example 10.13 of a noncompliant code is shown.

Noncompliant Code Example 10.13

```
union U1 // Noncompliant
{ float j;
int i;
}
```

Pre-processing Directives, Standard Libraries, and Identifiers

Rule R.20.4 (required): A macro shall not be defined with the same name as a keyword

The reserved identifiers, function, and macros that come under the standard library should not be defined or undefined or redefined. The names of the standard library macros, functions, and objects are prohibited to be reused. An example 10.14 illustration is discussed.

Noncompliant Code Example 10.14

```
# define sqrt cathy
```

In example 10.14, the library function *sqrt* is being redefined, such that wherever the keyword *sqrt* occurs in the code, the pre-processor is instructed to just replace the same with the value "cathy," where in a real case, the library function *sqrt* comes under the "math library" and is used to compute the square root of a given number.

Summary

The chapter dealt with the layered architecture of AUTOSAR which is agreed upon by the automotive companies. It dealt with a brief explanation on the software components in the architecture and the services provided. This chapter besides discussing AUTOSAR also discussed MISRA coding standards, the versions of MISRA, and the characteristics of MISRA. It also comprises a series of supporting examples for rules that are taken from various categories.

References

1. G.N. Vo, R. Lai, M. Garg, Building automotive software component within the AutoSAR environment – A case study, in *2009 Ninth International Conference on Quality Software, Jeju*, (2009), pp. 191–200. https://doi.org/10.1109/QSIC.2009.34

2. H. Bo, D. Hui, W. Dafang, Z. Guifan, Basic concepts on AUTOSAR development, in *2010 International Conference on Intelligent Computation Technology and Automation, Changsha*, (2010), pp. 871–873. https://doi.org/10.1109/ICICTA.2010.571
3. Q. Wang, B. Xin, C. Li, H. Chen, The realization of reusability in vehicular software development under AUTOSAR, in *2013 IEEE Conference Anthology, Anthology*, vol. 2013, (2013), pp. 1–6. https://doi.org/10.1109/ANTHOLOGY.2013.6785006
4. S. Martínez-Fernandez, C.P. Ayala, X. Franch, E.Y. Nakagawa, A survey on the benefits and drawbacks of AUTOSAR, in *2015 First International Workshop on Automotive Software Architecture (WASA)*, (Montreal, QC, 2015), pp. 19–26. https://doi.org/10.1145/2752489.2752493
5. AUTOSAR. Layered Software Architecture. 2017 Document ID 053: PDF file. 3 June 2020
6. AUTOSAR. Specification of I/O Hardware Abstraction. 2016 AUTOSAR CP Release 4.3.0 Document ID 047: PDF file. 3 June 2020.
7. AUTOSAR. Requirements on I/O Hardware Abstraction. 2017 AUTOSAR CP Release 4.3.1 Document ID 075: PDF file. 3 June 2020.
8. AUTOSAR. Utilization of Crypto Services. 2017 AUTOSAR CP Release 4.3.1 Document ID 602: PDF file. 3 June 2020.
9. AUTOSAR. Specification of Crypto Abstraction Library. 2014 AUTOSAR CP Release 4.1 Document ID 438: PDF file. 3 June 2020.
10. S. Fürst, M. Bechter, AUTOSAR for connected and autonomous vehicles: The AUTOSAR adaptive platform, in *2016 46th Annual IEEE/IFIP International Conference on Dependable Systems and Networks Workshop (DSN-W)*, (Toulouse, 2016), pp. 215–217. https://doi.org/10.1109/DSN-W.2016.24
11. Ward, David. (1998–2019). MISRA https://www.misra.org.uk. Accessed 1 Jul 2020
12. Jones, Nigel. (2002, July 1). Blog- Introduction to MISRA C *https://www.embedded.com/introduction-to-misra-c*. Accessed 1 Jul 2020
13. MISRA Compliance With Parasoft *https://www.parasoft.com/solutions/compliance/misra*. Accessed 1 Jul 2020
14. *http://automotive.roger.free.fr/articles/miscprev.pdf* [1999]. Accessed 1 Jul 2020
15. C static code analysis *https://rules.sonarsource.com/* [2008–2020]. Accessed 1 Jul 2020

Model-Based Automotive Software Development

K. Vinoth Kannan

Introduction

In the world of automobiles, the development of highly efficient and error-free software poses a great challenge. These challenges grow with the need of complex software algorithms and maintaining safety standards. To overcome these challenges, automobile software development companies are migrating from the conventional way of software development to Model-Based Software Development (MBSD). Software development model evolves based on process frameworks such as International Organization for Standardization/International Electrotechnical Commission (ISO/IEC). The need to analyze system behavior during integration led to MBSD [1]. MBSD is a visual method of solving complex embedded control systems.

Initially models are maintained as a high-level design diagrams, to discuss the functionality within different stakeholders like Software developers, Software architects, Domain and safety experts and, also to the customers. It is maintained as a skeleton for functions under development. When there is a request from OEM about comfort functions like Lane Keep Assist and Traffic Jam Assist, development activities are divided into different phases. In the initial phase of discussion, only the customer requirements are available to get a better overview and easy communication across different stakeholders like software architects who create high-level models. In late 2000, the Object Management Group (OMG) launched a model-driven architecture (MDA) that created a new paradigm in MBSD. The quality of the product enhanced by finding defects in earlier phase of development leads to a cost-effective approach with reduced cost in maintenance [2].

In general, models are a representation of vehicle function in abstract view. The importance of embedded software development increased in the last four

K. Vinoth Kannan (✉)
Robert Bosch Engineering Solution, Coimbatore, India

M. Kathiresh, R. Neelaveni (eds.), *Automotive Embedded Systems*,
EAI/Springer Innovations in Communication and Computing,
https://doi.org/10.1007/978-3-030-59897-6_4

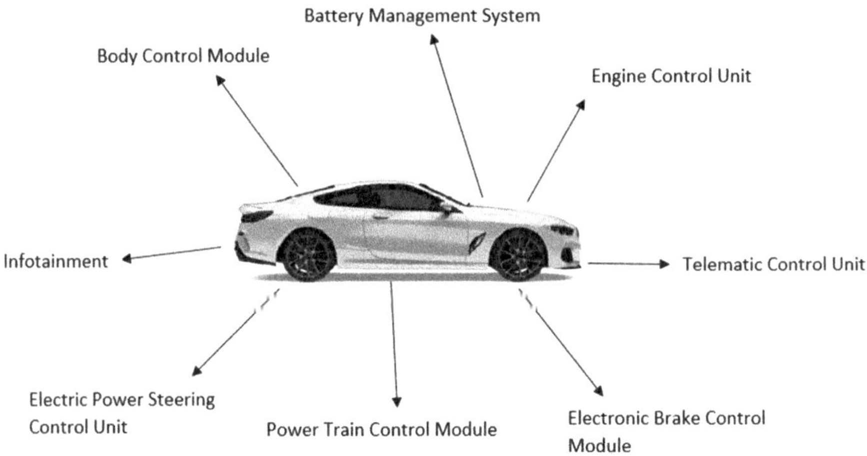

Fig. 1 Automotive electronic control units

decades. Critical products like TV, washing machine, and home automation are making human lives easier, but the primary goal for these kinds of product development is not safety. However, there are products which demand safety and robustness of the system, where human lives is the primary goal like aircrafts and cars [3].

Modern cars with comfort functions increase complexity of the system. Figure 1 visualizes various ECUs in a typical automobile (dcvizcayno.wordpress.com).

Software engineering is a systematic approach with different phases of development activities like requirement elucidation, architectural design, development, and testing. According to ISO/IEC/IEEE Standard 24765, software engineering is defined as "The application of a systematic, disciplined, quantifiable approach to the development, operation, and maintenance of software, that is, the application of engineering to software." In the modern software development, it is necessary to simulate the changes and verify the functionality of the system in the design phase itself to avoid software bugs in the production phase which has a huge impact on revenue of the Tier 1 supplier and OEM [4, 5].

In general, software is also one of the major reasons for accidents and product failure. A small change in ECU software of vehicle sold to customer leads to catastrophes. However, it is necessary to recall such safety critical products because it can be life-threating for humans as well as for the environment. Therefore, MBSD created a huge impact on automotive software development [4].

MBSD has become the most promising and widely accepted approach due to its features like auto code generation, verification and validation in the design phase, and less effort for domain experts to create a quick prototype of vehicle functions, allowing them to generate a code and verify it in the car to study the behavior. In the

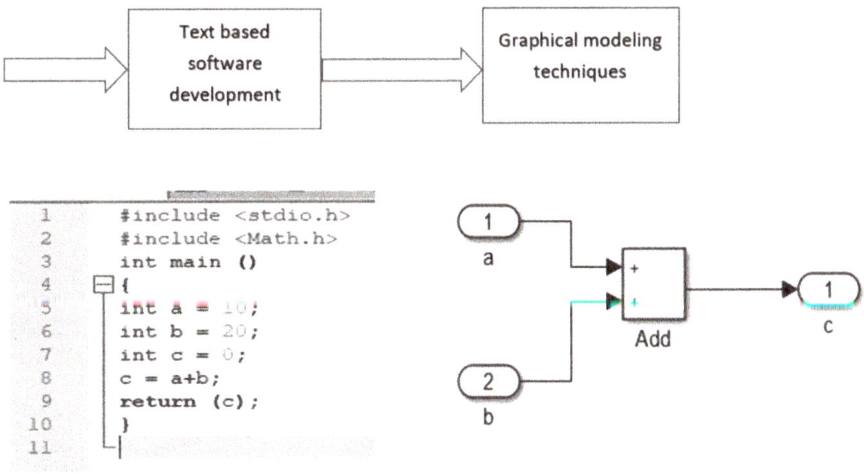

Fig. 2 Conventional software development vs MBSD approach

conventional software development process, the design information is documented in text-based formats. The automotive market always demands the product to be customizable and recyclable and to provide lifelong maintenance. In the conventional software design process, it is very difficult to understand and modify the manually written code. The difference between conventional and model-based approach is depicted in Fig. 2.

The key benefits of Model-Based Software Development include reducing the risk and cost by analyzing the model in the early phase of software development life cycle. Table 1 shows the achievement of leading OEM and Tier 1 suppliers [1]. As discussed earlier, MBSD improves the product quality and reduces the development time. The OEM demands the product to launch quickly into the market. So, MBSD is used for quick development and higher quality of the product. Maintenance of models developed using model-based development is highly cost-effective. It is also easily understandable across the project members. A graphical way of representation allows any developer to modify the implementation, thereby reducing the expert's dependency. It reduces the documentation costs [6].

The models are easily reusable by creating libraries without any extra effort. The library blocks used across the shared development environment will reduce the time and redundancy of the frequently used functions. Models are prepared as sub-blocks, hence easily replicable at the later stage of project. It is easy to analyze the error early and throughout the life cycle. In addition, assumptions are validated by using simulation blocks to match the system requirements. Auto code generation eliminates the coding errors, and the implementation is consistent to the given specifications. Extensive simulation reduces the debugging effort.

Table 1 Benefits of model-based software development

Company name	Tool used	Advantages
Airbus	SCADE and code generator	Reduction in errors by 20% and decrease in time to market
Eurocopter	SCADE and code generator	Cycle time reduction by 50%
GE and Lockheed Martin	ADI Beacon	Reduction in errors, cycle time, and cost
Schneider Electric	SCADE and code generator	8 times reduction of errors even when complexity increased 4 times
US Spaceware	MATRIXx	Decrease in cost and timeline for completion while subsequently reducing risks
PSA	SCADE and code generator	Cycle time reduction by 60% and 5 times error reduction
CSEE Transport	SCADE and code generator	Productivity increases from 20 to 300 SLOC per day
Honeywell Commercial Aviation Systems	MATLAB and Simulink	Increase in productivity. No errors in coding. Acquired FAA certification

Automotive Model-Based Software Development

In MBSD, software development life cycle (SDLC) is a process to produce high-quality and low-cost software with a short production time. It contains several phases like planning, design, implementation, testing, and deployment. Automotive software algorithms are more complex and supplied by different Tier 1 suppliers. To manage the complexity within the life cycle, there are many SDLC models created such as waterfall, agile software development, rapid prototyping, spiral, incremental, and synchronize and stabilize. The most commonly used software development life cycle in MBSD is V model which is shown in Fig. 3 [5].

In automotive industry, V model is the widely accepted one and also by the automotive safety standards like ISO 26262. Review checklist and quality documents are created for each phase. Entry/exit criteria are clearly defined for each phase.

• Requirement phase	Customer requirement gathered and system requirement derived in this phase
• Model-based design	Vehicle control algorithm designed by using MBSD tools like MATLAB/Simulink
• AutoCode generation	Once the model is validated against requirement, the code is generated out of model by using code generator such as RTW Embedded coder and Targetlink
• Integration	Once unit testing is performed, new feature is integrated into the software and once again is verified with other functions

V model is the extension of waterfall model also known as verification and validation model. In general, the left side of V model describes the development activities, and the right side is all about testing. It is important to finish one phase to move

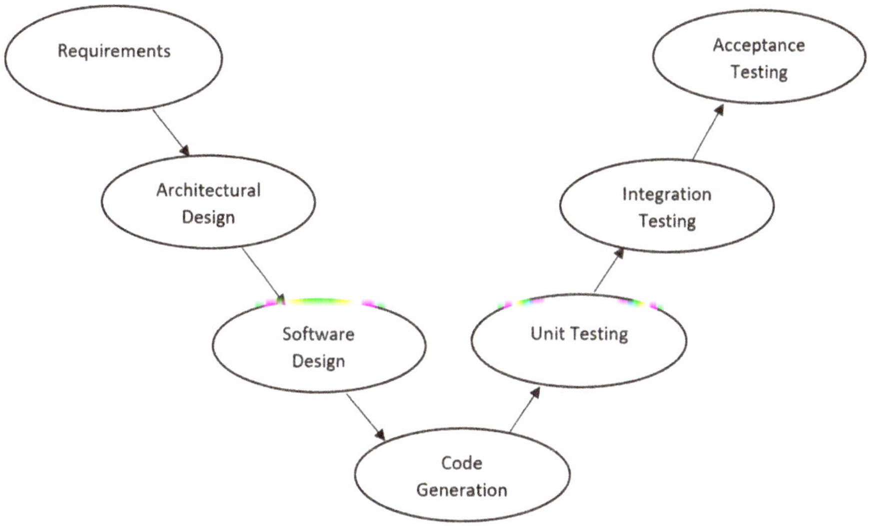

Fig. 3 V Model for MBSD

to another phase. Development and testing activities should happen in parallel to reduce the product development time. Some key benefits of V model are listed below:

- Testing starts in the early stage of software development, which avoids downward flow of error and makes it possible to find errors in an early stage
- Test specification prepared at the same time when the developer is creating the software
- Cost-effective as it avoids rework
- Testing happen in all stages of software development

Requirement Elicitation

Requirement elicitation or requirement gathering is a practice of discovering system requirements from different stakeholders. There are two types of most commonly accepted tools, namely, DOORS and Word for requirement elicitation and management. The DOORS tool provides a lot of options to link multiple requirements and provides traceability between customer requirements, unit level requirements and can also be linked with unit testing tools.

In automotive software development, requirements are classified as follows:

- Safety related
- Non-safety related

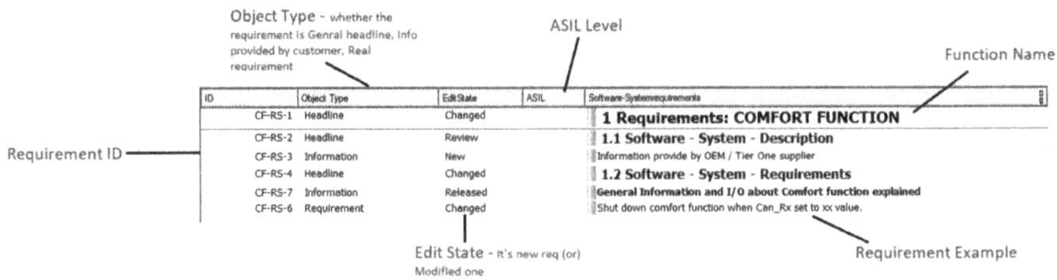

Fig. 4 Requirement analysis in DOORS

Safety-related requirements are treated with zero tolerance level in software development and testing activates. All safety-related test cases must be evaluated by system-level testing such as Hardware-in-Loop (HIL) and vehicle testing.

The process of requirement analysis using DOORS is depicted in Fig 4. Each requirement is filled with different attributes like object type, requirement ID, Automotive Safety Integrity Level (ASIL), and edit state. Object types are about the current state of requirement. Requirement ID usually starting with the function name is for traceability from software requirements specification (SRS) to design requirement, customer specification, and test specification. The requirement engineer generally writes SRS, and the software architect reviews it. ASIL is the important attribute for each requirement in automotive software development.

ASIL-related requirements are identified by applying simple filter options provided in DOORS. Safety tests and safety coverage are automated by creating simple scripts in DOORS.

Software Architectural Design

Software architecture (SA) is the basic blueprint creation for the entire system. In many OEM and Tire 1 suppliers, SA artifacts are created based on process standards ISO 26262 and ASPICE. Responsible software architects will create software architectural design documents and diagrams for the entire project, which are reviewed by another software architect. Software architectural design documents contain a black box diagram, a skeleton for vehicle function under development. Interaction between components of an ECU is shown in Fig. 5. Interaction between inter ECU and other ECUs is illustrated in a graphical way using tools like Enterprise Architect. Interaction is generally helpful for stakeholder discussion and issue analysis.

Design requirements are detailed design requirements created based on system requirement. Safety Artifacts like safety analysis is done for ASIL-related components. The software architect is responsible to ensure the functionality developed with necessary safety implementation recommended by ISO 26262. Safety engineers review the safety-related documents.

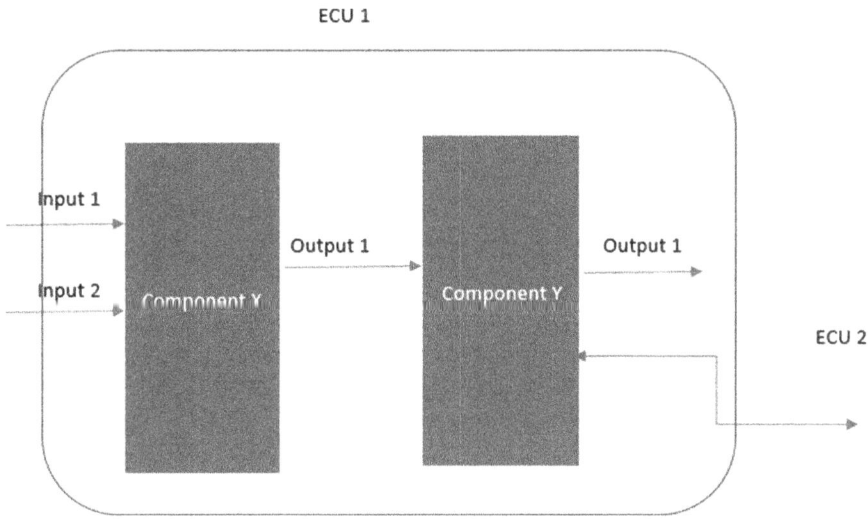

Fig. 5 Component interaction created using EA

In the recent trends, automotive vehicle systems are moving from mechanical to assisted electronic control units. In particular, a variety of highly automated driver assistance (HAD) or fully automated driver assistance (FAD) systems are being used. These systems influence the vehicle dynamics with high-risk factors [3], and the safety concerns are addressed by normative guidelines. The ISO 26262 standard addresses the international safety standards for automotive product development. Modern vehicle development should be in line with ASIL. ASIL is the risk classification of ISO 26262 compliance. Vehicle system protection in automotive software development also handles fail-safe conditions like hardware failure, environmental stress and software bugs which are analyzed with the help of ISO 26262.

Software Model-Based Software Development: MATLAB/Simulink

There are multiple MBSD tools and techniques available for automotive and aerospace software development [7]. MBSD is supported by modeling languages such as Systems Modeling Language (SysML) and Unified Modeling Language (UML) for efficient software development. The readily available tools for MBSD are MATLAB/Simulink, LabVIEW, ASCET, and Scade. Among these tools, MATLAB/Simulink is the most commonly used in vehicle function development and plant model development. MathWorks provides complete MBSD tool chain for modeling, code generation, and verification. Vehicle functions are modeled in a simulation environment and validated against plant model. MATLAB/Simulink offers the following benefits:

- Easy-to-build large model with modularity concept, like creating own reusable libraries
- Fixed-point model for timing-related real-time requirements
- Floating-point model for real vehicle system [plant model]
- Early validation of functionality and error detection
- Auto code generation
- Ensuring MISRA compliance by tools like Model Examiner (MXAM)
- Functionality testing by supporting tools like Time Partition Testing (TPT) and BTC Embedded Tester

Figure 6 shows the process of block creation using MATLAB/Simulink [8]. It is always time-saving, and users can create a reusable library which can be shared across the project member. Model referencing is a technique used to integrate a model from different developers as shown in Fig. 7 [8].

Sequential decision logics are designed in Stateflow environment. Time-triggered and event-triggered logics are modeled in Stateflow with the help of state transition charts and flow graphs. A simple "if else" Stateflow graphical pattern is shown in Fig. 8. The statements inside the square brackets are conditional statements, and the statements inside the curly brackets are action statements. The execution of this function starts from the default transition arrow. Upon reaching the junction, the arrow towards number "1" is checked or executed first.

Auto code generation is the main advantage of MBSD. It eliminates the manual coding effort and decreases the chance of error [9]. Real-time workshop, embedded coder, and dSPACE TargetLink are widely used to generate a code for the functions created by Simulink blocks. MATLAB provides various code generation options in different languages like C, C++, and hardware description language. It also

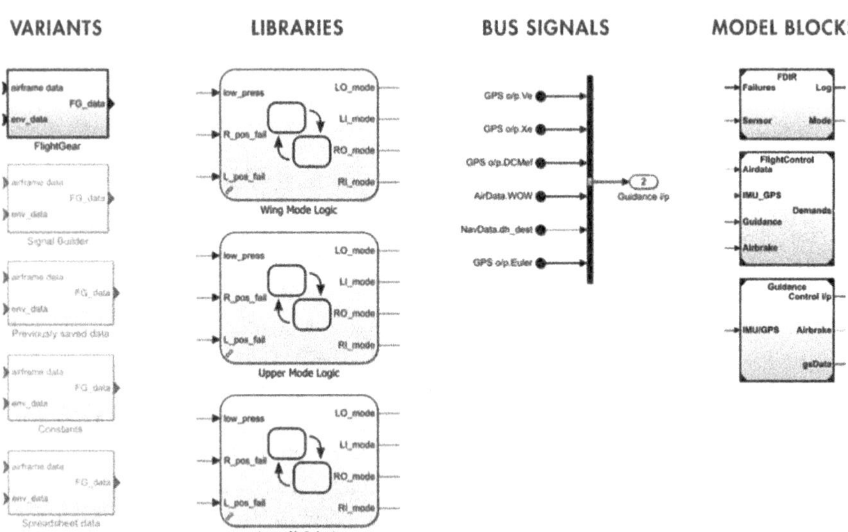

Fig. 6 Basic block creation using MATLAB/Simulink

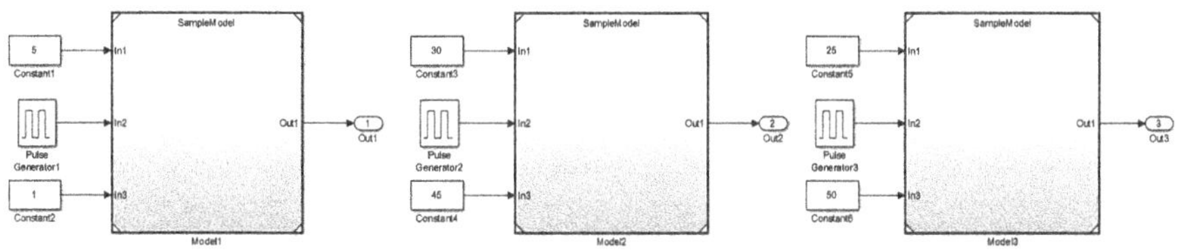

Fig. 7 Multiple components integrated based on model referencing

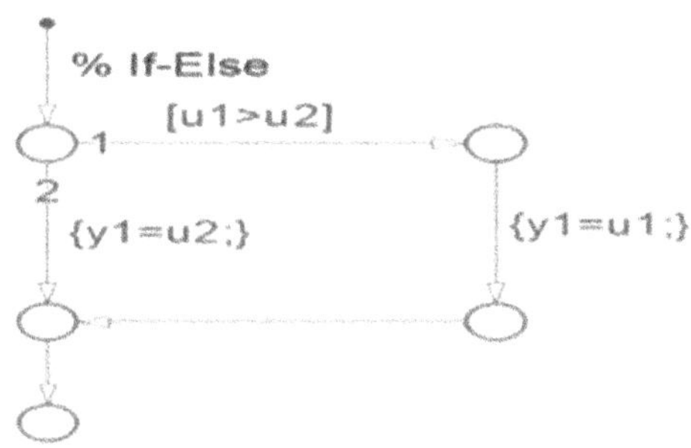

Fig. 8 Basic Stateflow illustration

complies with standards such as AUTOSAR, ISO 26262, DO-178, MISRA C, and CERT C. Since it is autogenerated, traceability between the requirements to code is also generated. Manual modification on the generated code is not recommended.

In MBSD, the algorithms are developed in a graphical way using modeling tools like Simulink/Stateflow as shown in Fig. 9 [10]. These models provide a simulation environment in all stages of software development from design to testing. In the design phase, physical models are derived based on customer requirement. In general, physical models are created in floating point.

In a digital hardware (microcontrollers), numbers are stored in binary words. In general, a binary word is a fixed-length sequence of bits (1s and 0s). Data types are defined by a sequence of 1s and 0s in order to interpret the software functions. Binary numbers are represented as either floating-point or fixed-point data types. Vehicle functions are converted to the fixed-point domain, and then it is described with Hardware Description Language (HDL). In HDL development process, it is necessary to provide the size of the variables and registers. The registers must be large enough to represent the value of parameters with the desired precision.

In floating-point algorithms, the operations are performed in large numbers (64 bits). It consumes more power and it requires more space. In order to avoid such problems, automotive algorithms are converted into fixed points.

Embedded coder is a code generator tool. As shown in Fig. 10, a code generated with embedded coder tool is compact readable C and C++ for mass production. It provides optimization options to improve code efficiency and allows integrating legacy code, tunable parameters, and data types. Embedded coder has built-in support to get compliance of MISRA C and AUTOSAR and provides code documentation and automated code verification to support standards like IEC 61508, ISO

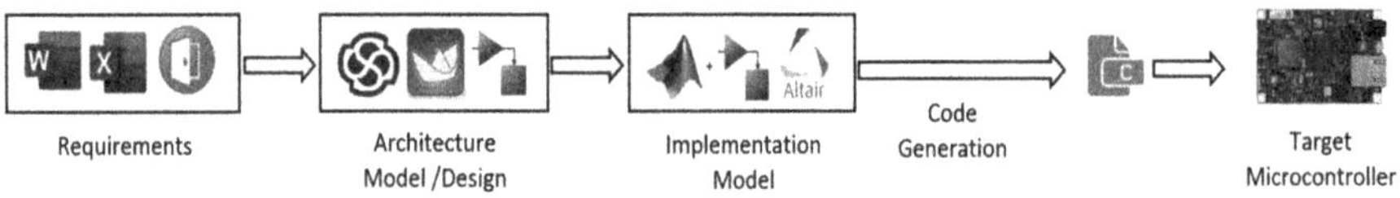

Fig. 9 Code generation using MBSD approach

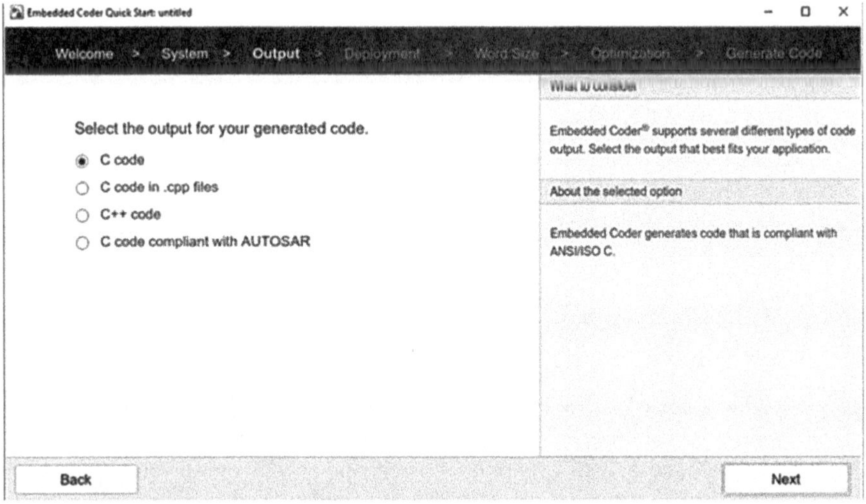

Fig. 10 Embedded coder options for output files

26262, and DO-178. To generate a fixed-point code, a fixed-point toolbox must be installed. This toolbox allows the user to change the properties of signals/parameter scaling, type, and offset.

Table 2 shows the different output file formats in MATLAB/Simulink. In addition to that, with the help of third-party tools, it is possible to generate a code specific to hardware including Intel, ARM, NXP, Texas Instruments, and STMicroelectronics. In general, it is easy to edit the properties of all Simulink blocks and libraries created out of Simulink/Stateflow. Figure 11 shows the Simulink Inport property window.

The model shown in Fig. 12 is a simple adder which takes two inputs, computes sum, and gives the sum as output. The code is generated from this model by setting the model configuration parameters to the required values. These parameters determine the target file (embedded real time or generic real time) and provides various options for customizing the code and the generated report. These options can also be left in its default state. After setting the model configuration parameters, the code is generated from the model by clicking the build icon.

The generated code is simulated with the help of Simulink block named S function. Once the code is generated, back-to-back testing is performed in the design phase itself. Vehicle function logics are verified in Model in Loop (**MIL**) mode by

Table 2 MATLAB/Simulink output targets

Software tools	Output targets
MATLAB	C/C++ executable MEX file C/C++ Static library
Simulink	Shared library (.dll) AUTOSAR interface Embedded real-time targets – ANSI/ISO C, C++

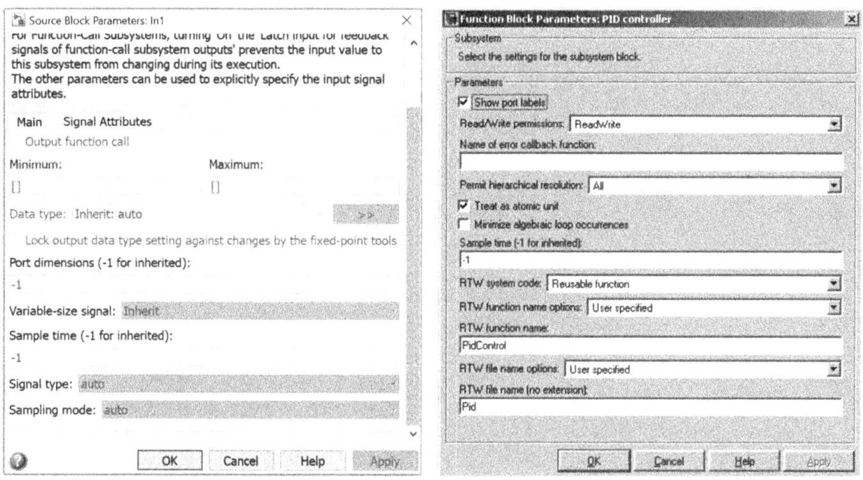

Fig. 11 Simulink Inport and Subsystem parameters

creating test cases using the Signal builder. The generated code is verified with the same test case in Software in Loop (**SIL**) mode.

S functions are compiled as a MEX (MATLAB executable) files. Legacy code integration is also possible with the help of S functions. Simulink Coder is used to generate a code out of S functions in a model. It is also possible to customize the generated code for S functions by writing a Target Language Compiler (TLC) file

TargetLink is a product from dSAPCE; it is very easy to generate code using TargetLink as options related to code generation are gathered in one place. The process of code generation in TargetLink is shown in Fig. 13 [11]. TargetLink's main dialog is responsible for code generation configuration. TargetLink is certified by TÜV SÜD (German certification authority) for safety-related software development according to ISO 26262 and IEC 61508. The Data Dictionary allows the user to modify elements such as scaling, data type, and variables.

In general, vehicle functions are modeled by Simulink blocks (or) blocks available in TL library as illustrated in Fig. 14. It is possible convert a Simulink model to TargetLink by a single push button named "Prepare system." In addition to that, it is possible to convert/generate a code out of a single subsystem provided each subsystem must have one function block.

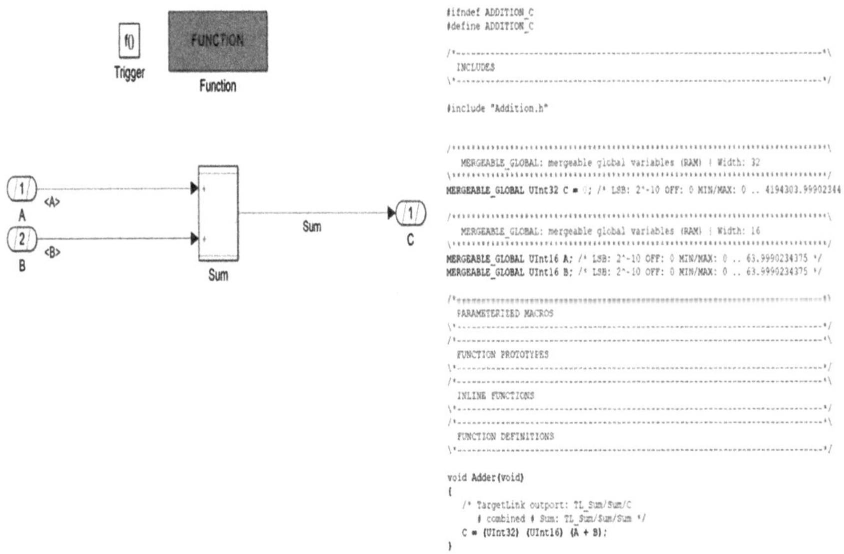

Fig. 12 Simple Simulink adder block and the corresponding generated code

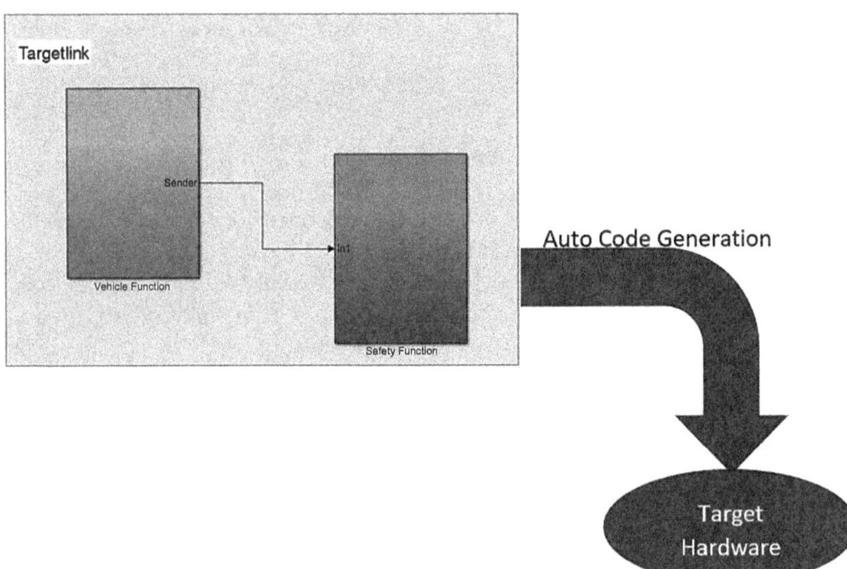

Fig. 13 TargetLink overview

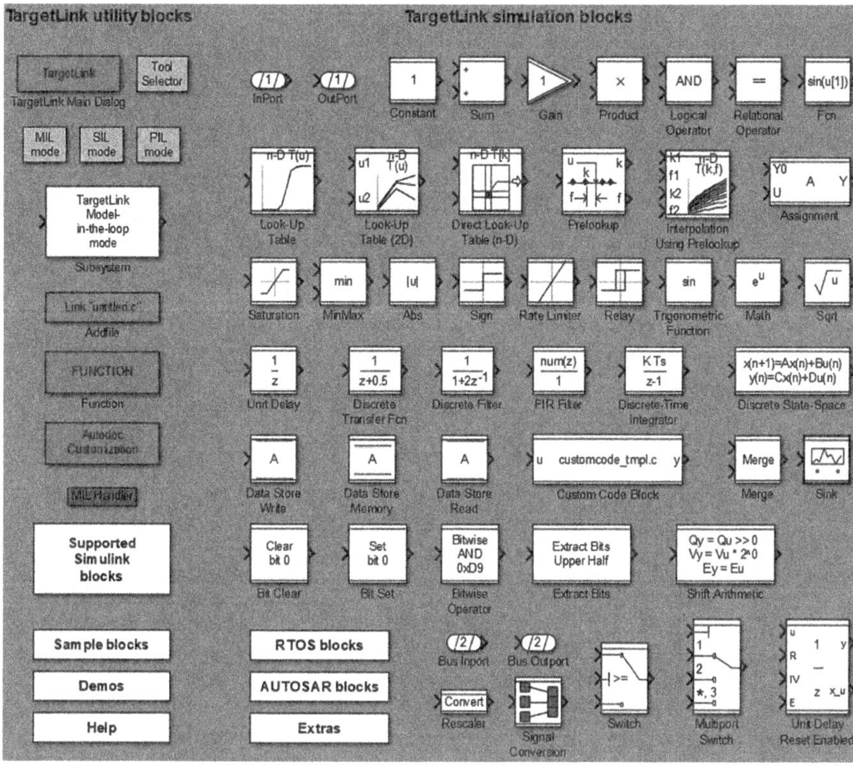

Fig. 14 TargetLink library blocks

A TargetLink-generated code is easily readable and provides traceability to requirement by providing comments in each subsystem. Figure 15 shows the TargetLink model and its corresponding generated code.

Simulation is a process of validating model and comparing results in various modes. TargetLink offers three kinds of simulation methods as illustrated in Fig. 16. By a single click, it is easy to change mode between MIL/SIL/PIL. New features or logics are validated against the real-time data collected from filed issues. It is very easy to introduce the real-time data into the simulation environment.

Unified Modelling Language in MBSD

Object-oriented design (OOD) is another approach widely used in automotive software development. UML uses the notations to describe OOD in MBSD. The UML diagram is the basis for software architectural design in most of the automotive industry. In general, programming languages are not in the level of abstraction to discuss about the design in the starting phase of product or software development.

UML is an open standard to provide higher-level abstraction and also facilitate the design discussion in the earlier stage of the project with different stakeholders and customer groups. UML is the basic blueprint where the designer provides detailed design to the programmer for implementation. In general, UML is a notation system, which helps to realize complex vehicle function algorithms in a graphical way.

UML diagrams can be broadly classified as structural diagrams and behavioral diagrams as the former one illustrates the structural aspects of a system and the latter one depicts the behavioral aspects of the systems with the actors and other systems. Figure 17 shows the different types of UML diagrams.

The following are the UML diagrams that capture and illustrate the static aspects of the systems:

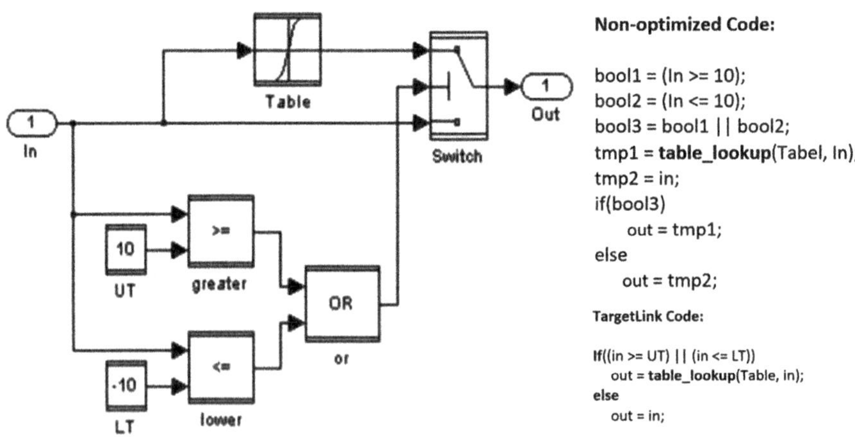

Fig. 15 Process of code generation

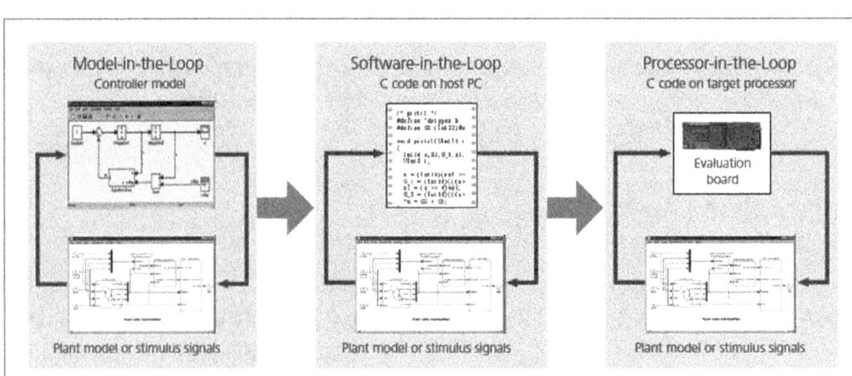

Fig. 16 TargetLink different simulation methods

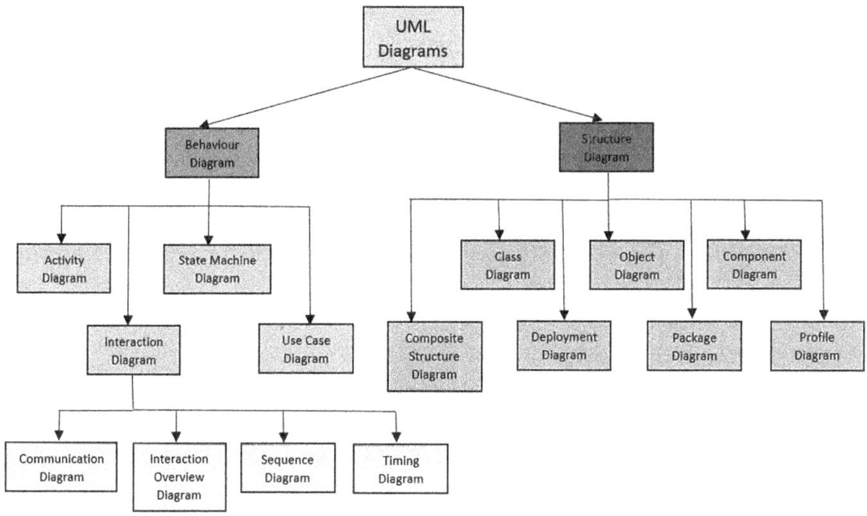

Fig. 17 Classification of UML diagrams

- A class diagram is the mostly widely used UML diagram that represents a system in an object-oriented manner.
- A object diagram is also used to represent the structural view of the system. Object diagrams are also similar to class diagrams in terms of usage. They are used to develop a prototype of the systems.
- A component diagram elucidates various components of a system and the interaction among them irrespective of time. Component diagrams provide an implementation view to the systems.
- A deployment diagram is one of the UML diagrams, which is used to model the physical entities of a system in the deployment view.

The dynamic aspects of a system are captured by behavioral diagrams. The following are the different types of UML diagrams that depict the behavioral aspects of the systems:

- A use case diagram represents various tasks that can be performed with the system. This also depicts the role of each actor in performing every use case of the system.
- An activity diagram depicts the flow of control in a framework. It comprises various activities and connections. The flow of activities can be consecutive, simultaneous, or even branched.
- A state chart diagram is used to convey the behavior of the systems when it is triggered by various events that happen within the system or outside the system. This is useful as most of the automotive systems are reactive to various external/internal stimuli.
- An interaction diagram is used to describe the flow of control and data. A sequence diagram is used to represent the sequence of activities/message opera-

tions that happen between various objects of a system with respect to time. A collaboration or communication diagram is used to visualize the structure of various objects involved in executing an activity of the system.

Vehicle function requirements are captured and then modeled at the system level. Modeling techniques involve two phases, namely, design phase and analysis phase. In the analysis phase, use case diagrams are used. A use case diagram is a user interaction diagram at the system level, where vehicle systems are described based on requirements. For each vehicle function, one or multiple use cases are identified, and clear tractability is established between the use cases and implementation (design requirements). In the design phase, UML diagrams like class diagrams, sequence diagrams, state diagrams, and deployment diagrams are used to visualize the core of system functionalities in a detailed way [12].

Summary

This chapter elucidates the Model-Based Software Development approach used particularly in automotive applications. As a vehicle's functional safety is considered to be one of the most important aspects of a car, a brief overview about that is given. Moreover, the chapter also discusses the various commonly used tools and techniques in MBSD approach. In particular, MATLAB/Simulink-based development and auto code generation techniques are explained in detail along with a brief introduction to Unified Modeling Language.

References

1. P.H. Feiler, D. de Niz, ASSIP study of real-time safety-critical embedded software-intensive system engineering practices, in *Special Report CMU/SEI-2008-SR-001*, (Carnegie Mellon University, Pittsburgh, 2008)
2. P. Mohagheghi, V. Dehlen, T. Neple, Definitions and approaches to model quality in model-based software development–A review of literature. Inform. Softw. Technol. **51**, 1646–1669 (2009)
3. S.M. Kugele, Model-based development of software-intensive automotive systems. (2012)
4. M. Jaskolka, V. Pantelic, J. Jaskolka, A. Schaap, L. Patcas, M. Lawford, A. Wassyng, *Software Engineering for Model-Based Development by Domain Experts* (Elsevier, Amsterdam, 2016). https://doi.org/10.1016/B978-0-12-803773-7.00003-6
5. International Organization for Standardization (IST/15). ISO/IEC/IEEE 24765:2010. (2010)
6. A. Bergmann, Benefits and drawbacks of model-based design. KMUTNB Int. J. Appl. Sci. Technol. **7**(3), 15–19 (2014)
7. M. Herrmannsdoerfer, T. Kofler, S. Merenda, D. Ratiu, J. Thyssen, Model-based development tools for embedded systems in the industry – Results from an empirical investigation, in *Software Engineering*, (2010)
8. Mathworks.com/Solutions

9. Nora Ajwad (2007), Evaluation of automatic code generation tools. in Lund University, Department of Automatic Control, MASTER THESIS, ISRNLUTFD2/TFRT-5793—SE
10. I. Fey, I. Stürmer, Code generation for safety-critical systems – Open questions and possible solutions. SAE Int. J. Passeng. Cars Electron. Electr. Syst. (2009). https://doi.org/10.4271/2008-01-0385
11. dSPACE.com, TargetLink product Information
12. W.-B. See, UML-based modeling approach for automotive system development, in *IEEE International Conference on Industrial Technology*, (Hong Kong, 2005), pp. 448–452. https://doi.org/10.1109/ICIT.2005.1600680

Vehicle Diagnostics Over Internet Protocol and Over-the-Air Updates

M. Kathiresh, R. Neelaveni, M. Adwin Benny, and B. Jeffrin Samuel Moses

Connected Cars

In the last two decades, the revolutionary progress in the field of automotive electronics and wireless technology and the ever-augmenting demands of the customers result in the automotive industry coming up with vehicles designed to sketch an incomparable driving experience from the traditional vehicles. By assisting this progress of meeting the demands of the clients like ensuring comfort, entertainment, connectivity, and other parameters during travel better than that of the pre-existing versions, cars of the recent times have truly taken the act of representing one's own individualism. Technologies like vehicle-to-vehicle communication (V2V) and vehicle-to-infrastructrue (V2I) communication, which enable communication between the automobiles and the environment through information exchange, not only come to aid in meeting the above said requirements but also enhance the user experience. This concept of vehicle interaction with the user and the surrounding environment is known as connected vehicles.

As shown in Fig. 1, the connected cars are automobiles that deploy several communication technologies to interact with the driver of the vehicle or with the other vehicles on the road or with the outside infrastructure. This vehicle interaction with the external world, resulting in the benefits of comfort, safety, connectivity, and entertainment, can be achieved using techniques like telematics, mechatronics, and artificial intelligence.

M. Kathiresh (✉) · R. Neelaveni
PSG College of Technology, Coimbatore, India

M. A. Benny
Infosys Limited, Bengaluru, India

B. J. S. Moses
Volvo Groups Private Limited, Bengaluru, India

M. Kathiresh, R. Neelaveni (eds.), *Automotive Embedded Systems*, EAI/Springer Innovations in Communication and Computing,
https://doi.org/10.1007/978-3-030-59897-6_5

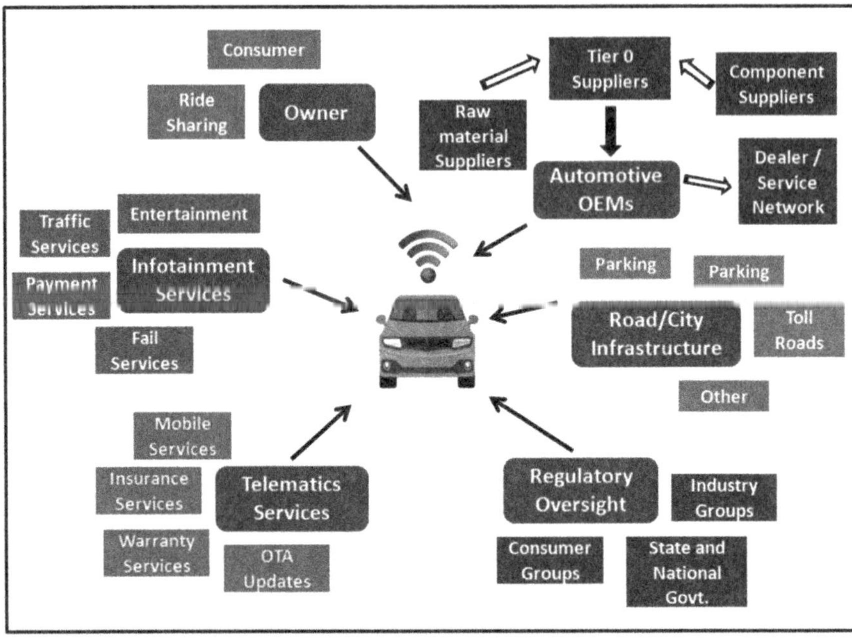

Fig. 1 Use cases of connected car technology

A typical connected car is built with intelligent features like muting the music on sensing the arrival of phone calls, so that the driver could indulge in without any disturbance; generating a signal of caution on the blockades or barricades or any obstructions that come in the way of the vehicle; stipulating information on an accident; providing the location of a nearby hospital or an emergency unit; transferring the vehicle status to the service providers enabling their access to remote monitoring; and many more additional features. Thus, cars of the recent times embedded with intelligence, enhanced versions of themselves with the aid of numerous sensors and actuators, have gained the ability to adapt themselves according to the current demands.

On-Board Diagnostics

From the earliest days of the commercial sale of the car, it has been obvious that maintenance and diagnosis are required to the proper working of the automobile. Until the primary 1970s, an honest deal of the routine maintenance and repair was done by car owners themselves, using inexpensive tools and equipment. Car owners cannot, as a matter in any case, do their own maintenance and repairs on certain automotive subsystems (particularly the engine). In fact, the quality shop manual used for years by technicians for repairing cars is rapidly becoming obsolete and is

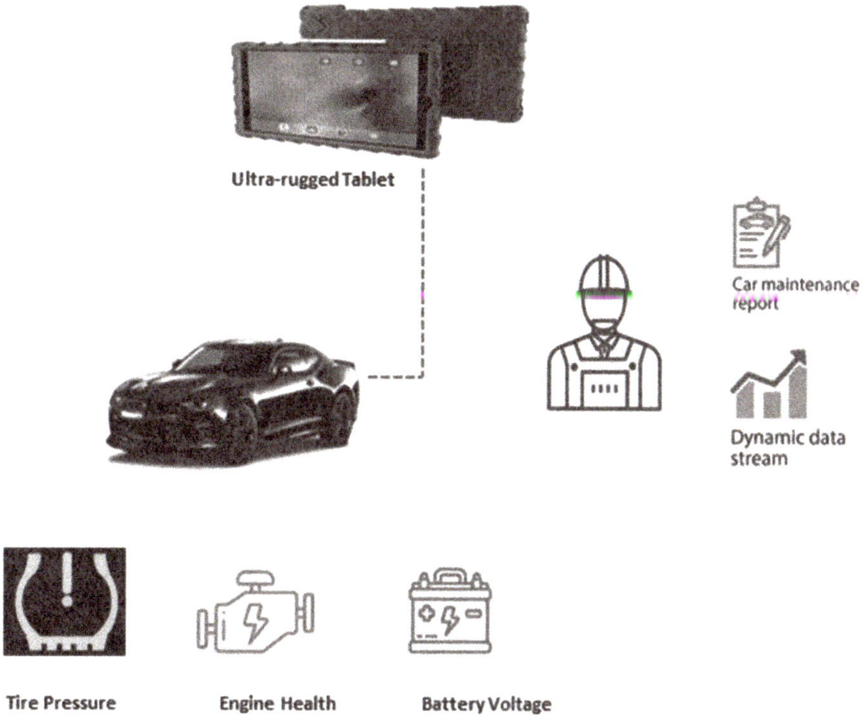

Fig. 2 Conventional vehicle diagnostics

being replaced by electronic technician aids. In the past two decades, the concept of on-board diagnostics (OBD) is used in vehicles. The OBD system has a malfunction indicator light present in the dashboard of the car to indicate any failure in vehicle parameters. The status of the vehicle is determined through various sensors, and in case of any issues, a unique code called Diagnostic Trouble Codes (DTC) is stored in the memory and later recovered by the technician using the scan tool to identify the cause of failure. Each of the DTCs has a letter and three numbers associated with it. The first letter is used to indicate the system related to the error, for example, "P" stands for powertrain which covers functions that include engine, transmission, and associated drivetrain accessories. "U" codes for network and vehicle integration which covers functions that are shared among computers and systems on the vehicle, etc. The numbers are manufacturer specific. The process of conventional vehicle diagnostics is shown in Fig. 2.

Diagnostics Over Internet Protocol (DoIP)

Even though the on-board diagnostics are useful to detect the errors, it requires the physical presence of the vehicle at the service center. The repair technicians can start the diagnostic analysis, only if the vehicle arrives at the workshop. The process performed by the repair technician to carry out analysis of the state of vehicle and fix the issues is very time-consuming. This in turn makes the vehicle owner inconvenienced. The remedy to these problems is to perform vehicle diagnostics remotely using the Internet. This reduces the overall downtime of the vehicle owner to fix the issues in the vehicle.

DoIP facilitates obtaining the data trouble codes and other status parameters from a vehicle remotely with the gateway module present inside automobiles [1, 2]. Using the gathered diagnostic information, the service person can do the major part of the repair work well in advance before bringing the vehicle to the service center. This helps the process of fixing the problems very quickly as the service technician can start the work immediately on getting the vehicle to the service center. DoIP can be used to continuously monitor the status of a vehicle to detect the faults present in it. This reduces the time for the detection of issues as well as the cost for the OEMs and also reduces the need to visit the service center. The process of conventional vehicle diagnostics is shown in Fig. 3.

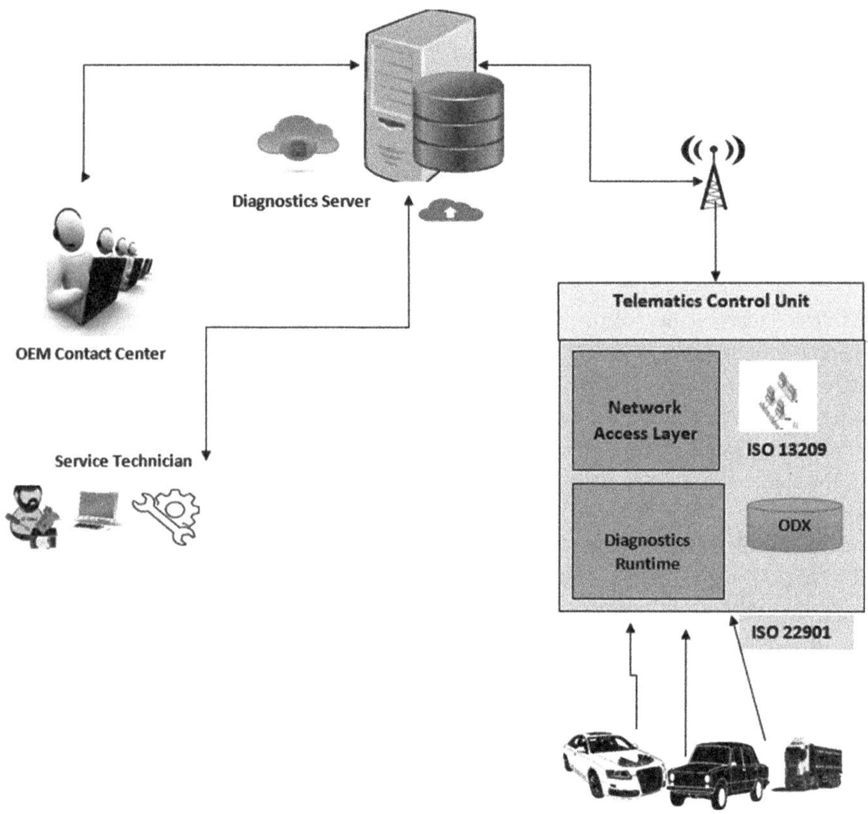

Fig. 3 Vehicle diagnostics over Internet protocol

Software Architecture in Automotive Electronic Control Units

The aspects of safety and comfort of the drivers are the important parameters under consideration. In order to ensure the appropriate levels of these parameters, numerous dedicated ECUs for anti-lock braking and engine management are deployed in today's cars. Sophisticated software application determines the overall functions that a typical ECU handles in a vehicle. Thus, the software of an ECU plays a vital role in determining the function of the system and facilitating installation of the software with the help of Gateway ECU present in vehicles. As shown in Fig. 4, the program memory for every ECU has two sections, namely, Bootloader and Application Area. The application section contains the programs which determine the tasks that the ECU needs to perform. On reset of ECU, the bootloader module executes, and the processor starts executing the application program. A bootloader module has got the following software components:

- Initialization module for hardware after power-ON and reset of the ECU
- Data download manager that facilitates programming flash memory blocks
- Unified Diagnostic Services (UDS) protocol stack for communication between the Target ECU and Gateway ECU
- Data Decryption Module based on a cryptography algorithm as an optional feature
- Data Decompression Module
- Application Layer

Functional domains in automobiles like powertrain, vehicle safety, infotainment, and chassis have many Electronic Control Units. Depending on the speed of data transfer required, different networking protocols like FlexRay, Media Oriented Systems Transport (MOST), Fast Controller Area Network (CAN), Local Interconnect Network (LIN), etc. are used to interconnect various ECUs. The Telematics Control Unit of a vehicle usually acts as a Gateway that facilitates interaction with external infrastructure and interactions among other ECUs.

Over-the-Air Updates

As there is a large number of Electronic Control Units present in a vehicle, the size of software application in a modern luxury vehicle can go beyond 100 million lines of program. It is quite obvious that the process of maintaining and updating such a complex software application for several thousand vehicles which are in production and millions which are in use is a tedious task. As shown in Fig. 5, in the conventional way of software update, the improvisations to the vehicles with the current software are provided by recalling them to the dealership either for the rectification of any problems in the software or in the circumstances of the addition of new features to the software, in order to enhance the performance of the vehicles. This

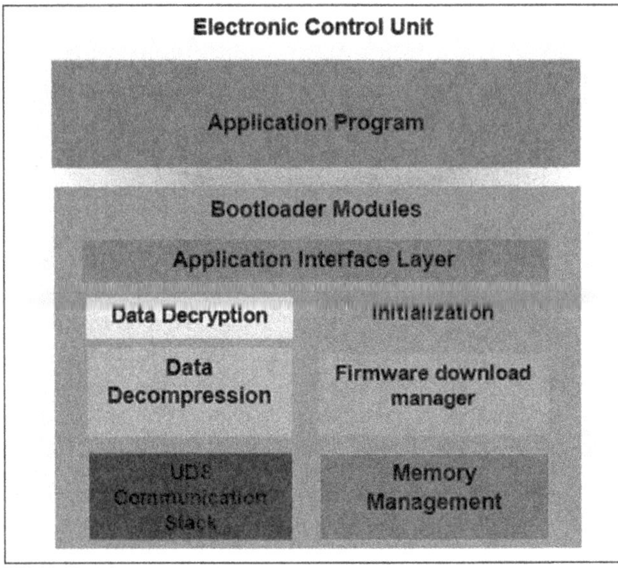

Fig. 4 Software architecture of an electronic control unit in vehicles

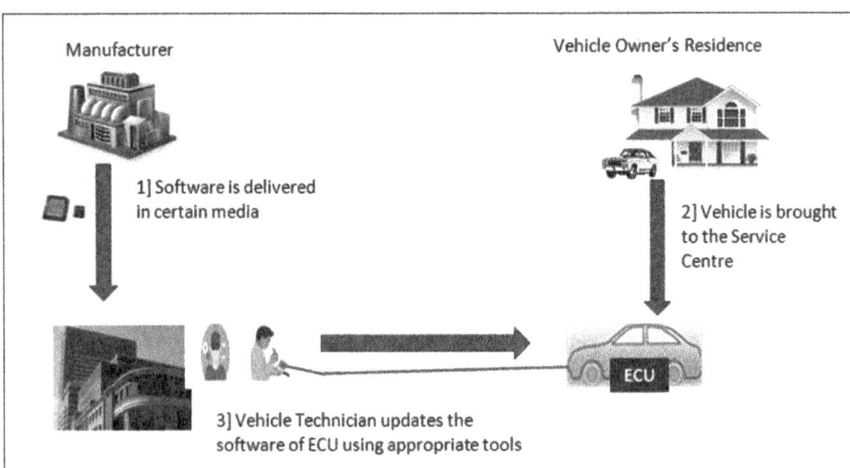

Fig. 5 Conventional Software Update Process for Automotive ECUs

process which demands the time and energy of the user may incur losses to the OEM gradually resulting in their downfall. So to prevent this, over-the-air update which enables the automatic update of the firmware of the ECUs emerges as a solution. Besides that, it also is an error-free and a faster mechanism that aids both programming and updating a car's firmware within its life cycle.

Over-the-air software updates refer to the practice of remotely updating the code on an embedded device [3]. The embedded hardware must be built with OTA functionality for this mechanism to work. As of now, OTA is a well-established concept for cellular phones, and its application to the automotive domain helps to handle sophisticated automotive systems. OTA uses a wireless medium to establish communication between OTA software server situated in the cloud and the telematics unit of a vehicle which acts a client. With the help of OTA process, it is possible to update software of vehicle ECUs at any location whether it is an assembly shop, location of a dealer, a service station, or the owner's parking area. It also does this software download as a background activity even when the vehicle is running, and once the download is done, it tells the service person or the user that it is ready for the installation of the updated software.

Figure 6 illustrates the process of updating the software in a typical vehicle ECU. OTA implementation is a three-step process as follows:

- Development of software updates to enhance the features of the automobile and storage of the newly developed software update files in the cloud database servers which is accessible to all the stakeholders
- Downloading of the newly received software update file into the memory module present in the Telematics Control Unit through the Internet
- Installation of the software update in vehicle ECUs

OTA is a collaborative process which requires participation from various stakeholders to give lifetime support to automobiles. The major stakeholders in the process of OTA are as follows:

- Tier 1 and Tier 2 suppliers who are responsible for development and supply of various automotive components
- The dealers who are responsible for the sales of vehicles
- Service stations for fixing faults and doing maintenance of vehicles

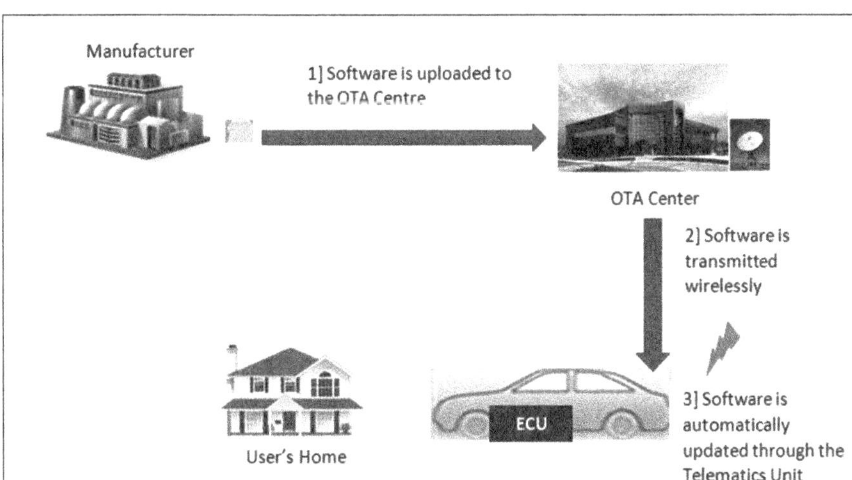

Fig. 6 Over-the-air software update process in automotive ECUs

- Owners of vehicles

Is the Process of DoIP and OTA Safe?

Anyone with a wireless communication interface can intercept critical data, which is a threat resulting due to the addition of these new features in the name of providing comfort to the users. This critical data when fallen into the wrong hands can be misused, thereby posing a potential threat. The resulting possibilities are that one can get hold of the code flashed into the ECU; tap into the vehicle's OBD interface; and potentially alter the transmitted code illicitly. By doing so, they can gain control over the car and cause catastrophes harming lives and the environment [4]. Therefore, it is significantly important enough to ensure the confidentiality, integrity, and authenticity of the data during the process of sharing over DoIP and OTA.

AUTomotive Open System Architecture (AUTOSAR) is an open community that provides software standards for automotive applications. For securing over-the-air updates, AUTOSAR suggests cryptographic algorithms [5]. Any cryptographic standard is considered vulnerable to brute force attacks. Cryptography alone will not be able to provide data security at the highest level; therefore an improved data security method which involves a combination of both cryptography and steganography is preferred to provide maximum security for over-the-air updates in automobiles [6].

The following are the preferred data security mechanisms that can be used to have reliable and safe vehicle diagnostics and software update through the Internet.

Cryptography

Cryptography is the art of encrypting sensitive information. Cryptography is based on mathematical theory and computer science practice. Cryptographic algorithms are designed with computational hardness assumptions, making such algorithms hard to break. Cryptography basically requires two steps: encryption and decryption. The encryption process uses a cipher in order to encrypt plaintext and turn it into cipher text. Decryption applies that same cipher to turn the cipher text back into plaintext. Cryptography is the development and creation of the mathematical algorithms used to encrypt and decrypt messages, and cryptanalysis is the science of analyzing and breaking encryption schemes. Cryptology is the term referring to the broad study of secret writing and bounds both cryptography and cryptanalysis.

There are five primary functions of cryptography:

- *Privacy/confidentiality*: Ensuring that no one can read the message except the intended receiver
- *Authentication*: The process of proving one's identity

- *Integrity*: Assuring the receiver that the received message has not been altered in any way from the original
- *Non-repudiation*: A mechanism to prove that the sender really sent this message
- *Key exchange*: The method by which crypto keys are shared between the sender and receiver

There are several ways of classifying cryptographic algorithms. They are categorized based on the number of keys that are employed for encryption and decryption and further defined by their application [7]:

- *Private key cryptography*: Private key cryptography or symmetric cryptography is the form of cryptography where only a single private key is used to encrypt and decrypt information as shown in Fig. 7. Using one common key creates a key management issue. It is possible that the private key may be stolen or leaked. Key management involves prevention of these risks by changing the encryption key often and appropriately distributing the key. Private key cryptography schemes are generally categorized as either stream ciphers or block ciphers. The most widely used private key cryptographic algorithms are Advanced Encryption Standard, Rivest Ciphers (RC1, RC2, RC3, RC4, RC5, and RC6), Data Encryption Standard, Twofish, Blowfish, and ChaCha.
- *Public key cryptography*: Public key cryptography or asymmetric cryptography is an encryption scheme that uses two mathematically related, but not identical keys—a public key and private key for encryption and decryption process. As shown in Fig. 8, the public key is used to encrypt, and the private key is used to decrypt. It is computationally infeasible to compute the private key based on the public key because of the one-way functions. Using this property the public key is shared freely, thereby facilitating the users to have easy and convenient encryption methods. Private keys have to be kept secret to ensure that only the owners of the private keys can decrypt the cipher text.

Public key cryptography ensures both confidentiality and integration of the original message. The one-way function can be realized using multiplication and factorization or by using exponentiation and logarithms. The ease of multiplication and exponentiation versus the relative difficulty of factoring and calculating logarithms,

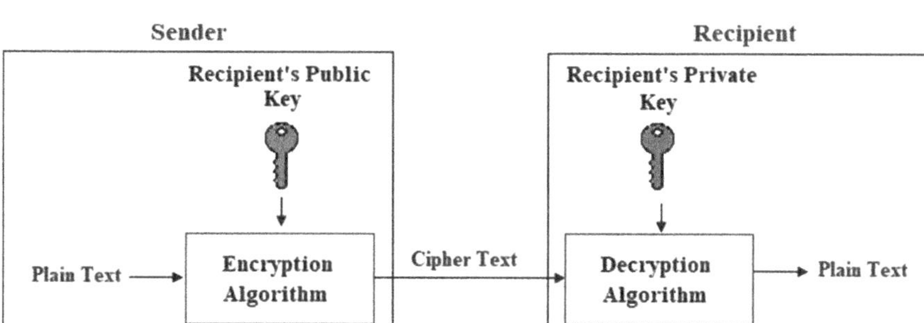

Fig. 7 Private key cryptography

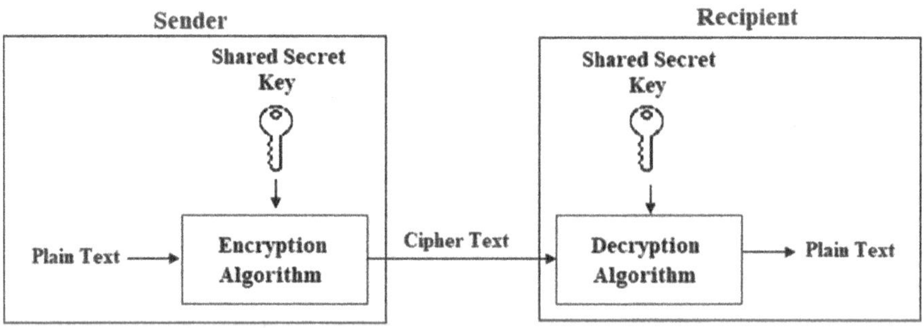

Fig. 8 Public key cryptography

respectively, create the one-way functions. The mathematical infeasibility in public key cryptography is brought by using a trapdoor in the one-way function so that the inverse calculation becomes easy with some prior information. Since public key cryptography is mathematically linked, the key generation process can be done locally at the receiver end. The most widely used public key cryptographic algorithms are Ronald Rivest, Adi Shamir, and Leonard Adleman (RSA) algorithm, Digital Signature Algorithm (DSA), Diffie-Hellman key exchange, ElGamal encryption, and Elliptic Curve Cryptography (ECC). Each of these algorithms has their own pros and cons; based on the application, a suitable algorithm is implemented for the cryptosystem.

Digital Signatures

A digital signature is the public key message authentication within the physical world. It is common to use handwritten signatures on handwritten or typed messages. They are wont to bind signatory to the message. Similarly, a digital signature could be a technique that binds an entity to the digital data. This binding will be independently verified by the receiver additionally as any third party. Digital signatures could be a cryptographic value that is calculated from the information and a secret key known only by the signer. The receiver of the messages needs assurance that the message received is the original message sent by the sender. This requirement is extremely crucial in business applications, since the likelihood of a dispute over exchanged data is extremely high. The most commonly used digital signature algorithms are RSA, ElGamal, DSA, ECDSA, EdDSA, Schnorr signature algorithm, and rapid digital signature. Each of these algorithms has their own pros and cons, based on the application.

Hash Functions

Hash functions use no key for encryption. Instead, a fixed-length hash value is computed based upon the plaintext that makes it impossible to retrieve the contents or length of the plaintext from the message digest. Hash algorithms are typically used to provide a digital fingerprint of a file's contents. Hash algorithms ensure integrity of the encrypted file. Hash functions are also commonly employed by many operating systems to encrypt passwords. Since hash functions produce a fixed-length value, there are a finite number of hashes for each type of algorithm. This makes collisions possible. A collision occurs when two different data (plaintext) produce the exact same hash. It's extremely rare for this to happen, but older hashing algorithms have encountered this problem. To overcome this problem, the hash value size has been increased for the same hashing algorithms. Some of the widely used hash algorithms are Message Digest (MD) algorithm which has different version like MD4 and MD5 with variable hash length, Secure Hash Algorithm (SHA), RACE Integrity Primitives Evaluation (RIPEMD), and Whirlpool.

Steganography

Steganography is the process of hiding secret data inside an ordinary, non-secret file or message to avoid detection [8]. Steganography can be used to conceal almost any type of digital content, including text, image, video, or audio content. The content to be concealed through steganography is often encrypted before being incorporated into the cover file. The use of steganography can be combined with encryption as an extra step for hiding or protecting data. Steganography is generally used to convey a secret message or code. There are many legitimate uses for steganography, but malware developers have also been using steganography to obscure the transmission of malicious code. Image steganography is the most commonly used steganography method to embed the secret data. In image steganography, the input message is inserted into the cover medium using special algorithms. Based upon the domain type, image steganography techniques are classified as spatial domain and transform domain techniques.

Summary

Tremendous growth in the field of automotive electronics, wireless communication, and information technology has made the processes of diagnostics over Internet protocol and software update over the Internet possible. As these processes use the Internet, an unreliable channel for information exchange, this involves lots of risks in terms of security of the data. AUTOSAR specifies cryptographic algorithms to have secured data transmission in automotive applications. It is also true that the cryptographic algorithms when combined with the art of steganography provide an additional security for the data being over the Internet during the process of DoIP and OTA.

References

1. M. Johanson, P. Dahle, A. Soderberg, Remote vehicle diagnostics over the Internet using the DoIP protocol, in *Proceedings of the Sixth International Conference on Systems and Networks Communications*, (IARIA, Barcelona, Spain, 2011), pp. 226–231
2. J. Lee, E. Lee, S. Park, Extended communication interface for remote vehicle diagnosis using Internet protocol, in *Proceedings of the 19th Asia-Pacific Conference on Communications (APCC)*, (Denpasar, 2013), pp. 421–426
3. M. Shavit, A. Gryc, R. Miucic, Firmware update over the air (FOTA) for automotive industry. SAE Technical Paper, in *Proc. of Asia Pacific Automotive Engineering Conference*, vol. 14, (2007)
4. H. Yu, C.-W. Lin, Security concerns for automotive communication and software architecture, in *Proceedings of IEEE Conference on Computer Communications Workshops*, (San Francisco, CA, 2016), pp. 600–603
5. AUTOSAR. Specification of crypto interface for adaptive platform AUTOSAR AP release 17-10. Document ID 883, October 27 (2017)
6. K. Mayilsamy, N. Ramachandran, V.S. Raj, An integrated approach for data security in vehicle diagnostics over IP and software update over the air. Comp. Elect. Eng. **71**, 578–593 (2018)
7. N.B.F. Silva, D.F. Pigatto, P.S. Martins, K.R.L.J.C. Branco, Case studies of performance evaluation of asymmetric and symmetric cryptographic algorithms for embedded systems. J. Netw. Comp. Appl. **60**, 130–143 (2016)
8. M. Hussain, A.W.A. Wahab, N.B. Anuar, R. Salleh, R.M. Noor, Pixel value differencing steganography techniques: Analysis and open challenge, in *Proc. of IEEE International Conference on Consumer Electronics-Taiwan (ICCE-TW)*, (2015), pp. 978–979

Automotive Cybersecurity

Ashish Jadhav

Introduction

Automobiles of today and of the future are a part of a well-connected network of vehicles which access multiple communication and informational services over cloud and across peers. The electrical infrastructure of a vehicle is made up of a large number of ECUs [electronic control units] communicating over multiple intra-vehicular networks. These are also connected through the in-vehicle infotainment [IVI] and telematics sub-systems to the external world over Internet-based protocols. As such the automotive world is exposed to security and privacy exploits prevalent over the cyberspace.

Automotive cybersecurity is a study of the issues of security of the automobile system and the privacy of the automobile user information in this well-connected vehicular scenario. Compared to any other consumer electronic systems like PCs, mobiles, and IoT devices, a vehicle has its own characteristics in terms of its architecture, usage, and upgrade constraints. In fact Internet of Automotive Things [IoAT] is the term that is used in the vehicular domain of things. As such classical security and privacy techniques like vulnerability and threat analysis, cryptographic security, privacy lists, Public Key Infrastructure [PKI], digital certificates, security updates, etc. need to be relooked and redesigned for the automotive systems.

In recent times various automotive vulnerabilities have been utilized, and security hacks of vehicles have been reported. An additional dimension is that an automobile is a safety critical system and the implication of such hacks in addition to the financial or time loss could also be on the life of the automobile users in the vicinity of a compromised vehicle or the pedestrians nearby. Hence providing security to an automobile user is a very important and challenging area. Automobile OEMs [Original Equipment Manufacturer] are already announcing plans for driverless

A. Jadhav (✉)
Ramrao Adik Institute of Technology, Mumbai, India

M. Kathiresh, R. Neelaveni (eds.), *Automotive Embedded Systems*, EAI/Springer Innovations in Communication and Computing,
https://doi.org/10.1007/978-3-030-59897-6_6

cars and ADAS [Advanced Driver Assistance System]. Such systems also would need innovative approaches for their security.

In this chapter, we begin with a description of the general principles of cybersecurity. Then we see some aspects of how automotive systems have evolved which helps us in understanding how there are differences between traditional cybersecurity and automotive cybersecurity. Next we look at the automotive cybersecurity threats and the in-vehicle infotainment system that is a crucial component from an automotive cybersecurity perspective. Next we look at the three very important standards in the automotive world from a cybersecurity view. These are AUTOSAR, ISO 26262, and ISO/SAE 21434. Finally we conclude this chapter with some very upcoming initiatives on the use of blockchains in the automotive industry and a brief about the MOBI standard.

Cybersecurity

With the evolution of computer systems, computer networks, computer programs and data, the field of computer security, network security, software security, and data security have undergone a revolution in the past few years. Today, we have computing devices which are ubiquitous, always on and always connected to the Internet. Computing systems today are in the form of mobile phones, tablets, and embedded systems in various things – Internet of Things (IoT). Their typical characteristics are huge volumes, low power consumption, and low hardware capabilities compared to a dedicated computing system like a personal computer or a server. Due to the proliferation of cheap mobile communication, the mobile computer network has innumerable devices used for various applications. Software is truly distributed across multiple devices and is also available as a service through the cloud. This is the world of information and data explosion. There are countless intelligent applications deployed which are processing information across multiple domains.

Cybersecurity deals with security of today's countless devices connected over the Internet executing distributed information processing applications to provide services to today's mobile users and things. It deals with security of the information on the computing devices with end user, cloud and embedded in things. It also deals with security of information in transit and in storage through the network. It deals with the security of flow of information from the end user to the application service provider through various intermediate service providers like hosting services, storage services, network services, etc. Cybersecurity is concerned with the procedures, processes, tools, techniques, and infrastructure required to enable security and privacy in today's world.

Cybersecurity can be further sub-divided into the following security domains from a decomposition perspective:

1. Application security: Dealing with the security of various applications providing services, user interfaces, and application interfaces which can be compromised

2. Information security: Dealing with the security of information and data that is entered, generated, measured, stored, transferred, and displayed by the application
3. Network security: Is the security of the network consisting of hardware and software components that are utilized by the distributed application for the transfer of information and data across geographically distributed application components
4. Computing system security: Is the security of the computing systems that could be personal computers, laptops, mobile phones, tablets, devices, things, etc. which are responsible for execution of the distributed application

Though we can view an application providing some service to a user, decomposition into the abovementioned security entities is important from a perspective of study, design, implementation, testing, maintaining, and upgrading a system with the objective of fulfilling security-related requirements and goals.

Threats to cybersecurity are caused when security aspects like confidentiality, integrity, and availability of system or data are compromised by a malicious user. Confidentiality of data implies that only the intended user can access the data. Integrity deals with the property that the data has not been tampered or changed by an adversary. Availability of the system and data deals with providing the legitimate users access to the system and data.

The objectives of cybersecurity are to ensure confidentiality, integrity, and availability to the users of the application or system. Attacks and hacks are carried out by adversaries to compromise the confidentiality, integrity, or availability of a system or application by exploiting various vulnerabilities present in the system or the application. These could be either in the system/application or in the information database or the network or the systems hosting the application.

This could lead to cybercrimes committed to genuine system users, to cyberattacks which are planned operations with a motive to compromise the system or data for some intention, or to more serious cyberterrorism which is used to cause unrest and panic by compromising security of the applications and data.

Cybercriminals use malicious code (malware) to exploit vulnerabilities and compromise systems and steal data. Some common types of malware include virus, Trojan, ransomware, adware, and botnets. Some cyberattack techniques include SQL injection, phishing, man-in-the middle attack, denial-of-service attack, etc.

Some cybersecurity techniques to protect systems from cyberattacks are as follows:

1. Employing cryptographic techniques for system and application security. Recently quantum computing has been demonstrated, and quantum cryptography is also an upcoming technology that could have a major impact on the implementation of cybersecurity in future systems.
2. Using secure protocols for communication and storage of data.
3. Regularly updating operating systems and application software with the security fixes.
4. Having a strong antivirus software and keeping it up-to-date.
5. Having strong passwords and regularly changing them.

6. Avoid opening emails and clicking on links and executables from unknown sources.
7. Avoid connecting computing systems to public insecure Internet connections.

To ensure that the goals of cybersecurity are satisfied in a system or application, cybersecurity has to be included as an integral part of the software development and deployment process for the application:

1. Cybersecurity requirements are to be included in addition to the functional requirements of the system/application.
2. Cybersecurity considerations to be considered in the system architecture design.
3. Cybersecurity considerations to be included in the coding of the software and the design of the data bases.
4. Cybersecurity testing, verification, and validation activities to be carried out. At times penetration testing and ethical hacking too needs to be performed to ascertain the security of the developed system or the application.

Automotive System Evolution

Automobiles have gone through a major transformation from mechanical machines to electronics vehicles. Olden vehicles had very few electrical parts for starting, ignition, headlights, and blinkers. The earlier vehicles had a very simple electrical battery-driven supplementary system to support the internal combustion engine as the primary locomotive power in the vehicle. Later some sensors and cluster meters used electronics followed by the tuner/radio and audio systems. From those elementary electronics systems in olden day cars, today's vehicles have hundreds of ECUs interconnected with multiple communication networks. Due to the complexity of the implementation today, it is popular to use multicore ECUs [1]. The tuner has been transformed into a complex in-vehicle infotainment system which has diverse communication capabilities and interconnectivity.

Today, cars run multiple applications which enable the vehicle to be networked to other cars or with the infrastructure. As part of these information processing applications, it is necessary for the vehicle to communicate over the Internet. The automotive system can be further sub-divided into the following sub-systems:

1. Body and chassis
2. Powertrain
3. Engine
4. Climate control
5. Braking
6. Steering
7. Exhaust
8. Infotainment and cluster

All the sub-systems have ECUs, software components, and communication abilities to interact with other vehicular sub-systems.

The amount of electronics and software in cars having advanced safety features and autonomous driving capabilities is high to enable drive-by-wire capabilities. In fact modern cars are very complex real-time embedded software-controlled and information-driven mechanical systems. More than 50% of the cost of today's cars is of electronics and the software that goes in it. Across the multiple ECUs, there could be more than 100 million lines of code. Today's cars are truly software driven.

A car is a safety-critical system, and the mechanical, electronics, and software components together have to satisfy the safety goals required for the vehicle. The electronics equipment and all the systems in the car are designed to operate under severe environmental conditions like cold, hot, and dusty situations.

Typically the sub-systems of an automotive system are designed to have a working life-span of greater than 10 years. The infotainment system is typically built to interface with mobile devices and external interfaces. These typically evolve at a faster rate. These are consumer electronic devices and are typically replaced in 2–5 years. Integrating such diverse systems and providing applications in a vehicle which are safety-critical is one of the biggest challenges of automotive system design.

Automotive Cybersecurity

Consider the diagram given in Fig. 1, which shows the block diagram of a connected vehicular network.

Compared to a computer network or a mobile ad hoc network, there is a major difference when we consider the vehicular networks. Whereas the node in a computer or mobile network is typically a computing device which is connected, it is much more complex in the case of vehicles. The vehicle which is connected to the

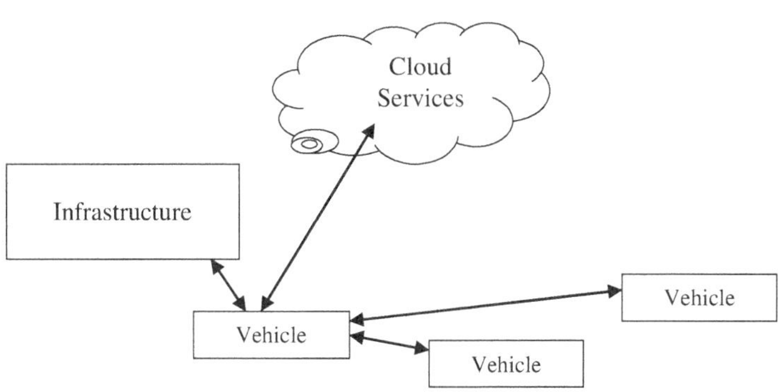

Fig. 1 Connected vehicular network

network is in turn consisting of hundreds of ECUs interconnected by diverse networks like the following:

1. Control Area Network [CAN] typically used for powertrain, engine control, etc.
2. Local Interconnect Network [LIN] typically used for powered windows, seat belts, climate control, door locks, etc.
3. Media Oriented Systems Transport [MOST] typically used for multimedia communication and infotainment systems
4. FlexRay typically used for steer-by-wire, brake-by-wire, etc.
5. Ethernet typically used for telematics applications
6. Dedicated Short-Range Communication [DSRC] for vehicle-to-vehicle and vehicle-to-infrastructure communication, etc.

Most of the ECUs are running real-time critical applications, and a failure or malfunction can be fatal in this safety-critical system. Typically, a secure gateway device is used for interconnecting these diverse networks, and a firewall is used for the Ethernet connection to the external world. Automotive systems can also have intrusion detection systems [IDS] to detect anomalous activity in the network. A diagram of these systems is shown in Fig. 2.

One more challenge from a cybersecurity perspective in an automotive system is that of software updates. While we are used to have automated updates and planned updates in computing systems and mobile systems for fixing security loopholes, updating software in a vehicle is a challenge [2] due to the time-critical nature of the systems and the safety and availability requirements.

Another consideration which cannot be taken lightly in terms of an automotive system is that we cannot say that a system is 100% secure. The same is true about safety, we cannot say that a system is 100% safe. There have been reported cases in the past of accidents that have been caused due to safety failures. Safety processes have matured over the years, and by using techniques like redundancy in the system, the chances of failures can be reduced to satisfy the practical purpose of system

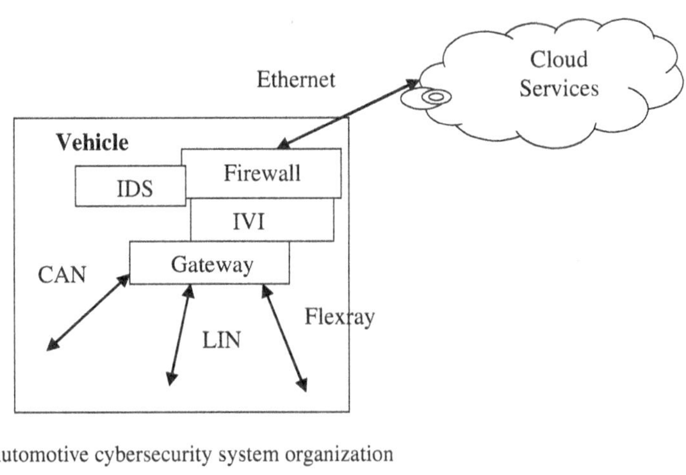

Fig. 2 Automotive cybersecurity system organization

usage. Also, a safety failure is due to some component failure, misbehavior, or environmental factors that can be either predicted or analyzed. The difference in security is that the accidents caused by security breaches are intentional and are cybercrimes. They are caused by an intelligent adversary who hacks into the system and is responsible for the failure caused to the system which could in turn lead even to a safety failure. So indirectly the safety of the system has been manipulated by an intelligent adversary.

There is one more important difference in an automotive system from the point of view of implementation of cryptography to provide cybersecurity. Security heavily relies on cryptographic algorithms. The cryptographic algorithms that implement cybersecurity are basically open algorithms in the sense that the algorithm is publicly known to everyone. The strength of the algorithm primarily depends on the cryptographic key that is used by the algorithm. The symmetric key has to be kept secret in case of a symmetric key-based cryptography between the two parties exchanging secure information. The private key has to be kept secret in case of asymmetric key-based cryptography. Usually the secret key is associated with a user in case of a computational system like PC or mobile and is protected by the password or passphrase which is known and entered by the user for cryptographic operations. In an automotive scenario, this is difficult. As an automotive system is in turn a distributed network of ECUs, the sub-systems would have to communicate securely without human intervention. It is also not possible for the user to enter the password while simultaneously driving the vehicle. The vehicle has to have built-in mechanism to store the secret key and use it without human intervention. So to ensure this secrecy in automobiles, hardware-based mechanisms like a Hardware Security Module [HSM] or on-chip security implementation are used. If a hacker tries to hack the secret key from the hardware module or the on-chip store, the hardware is destructed and fails and has to usually be replaced.

Ethical hackers have taken full control of vehicles in the past and have demonstrated the importance of automotive cybersecurity.

There are a plethora of SAE, ISO, AUTOSAR, and IEEE standards which deal with security of various aspects and components of an automotive system right from ECUs, vehicular networks, to communication protocols. Of special interest is the ISO/SAE 21434 joint standard for cybersecurity of automotive systems, and we will consider this in details in a later section.

Automotive Cybersecurity Threats

The three major threats that exist in any connected vehicle system are threats dealing with the following:

1. Confidentiality—An attacker can gain access to the confidential vehicular data. Data could be belonging to the vehicular driver, connected phone, vehicular data, or OEM-related data.

2. Integrity—An attacker can modify the data in the vehicle. This could be programs in the ECUs. Malicious code can be introduced.
3. Availability—An attacker can prevent the data from the vehicle to be accessed by a legitimate connected external vehicle or a user.

The threat of hacking a car could lead to the following scenarios:

1. Injecting malicious code in the vehicle
2. Getting remote access and taking over the controls of a vehicle, i.e., the driver or the controller of the vehicle is unable to control the vehicle
3. Theft of information flowing from the vehicle to another vehicle, to the infrastructural computer, or to the cloud
4. Getting access to the back-end systems of the OEM or service providers of the vehicle

Traditional security uses a connected system to frequently update security patches. Automotive systems unlike consumer electronics systems have certain challenges and complications in the application of software updates:

1. Automotive system might not be always connected. There could be regions and durations where connectivity does not exist.
2. Difficulty of updating software in automotive systems. The system cannot take time to update, and it needs to start immediately. It cannot start an update when it is being driven as it is a real-time safety-critical system.
3. Unlike a PC or a mobile phone, an automotive system could have software scattered into multiple ECUs and processors connected over diverse networks. Handling distributed updates across the system is complex.

The major threats to a vehicle from a cybersecurity perspective are the following:

1. Brought-in devices. Mobile phones, tablets, and computers which have their own connectivity and which are in turn connected to any of the vehicular system.
2. Vehicular wireless connections like Bluetooth and Wi-Fi, required for infotainment and telematics applications. These could include vehicle-to-vehicle, vehicle-to-infrastructure, or vehicle-to-cloud connectivity.
3. Physical thefts of a chip or module from the vehicle.

A popular threat model in use is briefly described here. The Microsoft STRIDE model [3] classifies potential threats according to the types of exploits that are used:

1. Spoofing—is a threat on the authentication and involves impersonating a legitimate user
2. Tampering—is a threat on the integrity of data and involves modification of data or code by an unauthorized entity
3. Repudiation—is an threat to non-repudiation and involves improper claim on performing some actions
4. Information disclosure—is a threat on the confidentiality of information and involves unauthorized access to information

5. Denial of service—is a threat to the availability of the services to the legitimate users
6. Elevation of privilege—is a threat to authorization and involves illegitimate ways to get authority

For an automotive system threat modeling, the STRIDE threats are considered against each component of the system that is exposed as a threat surface, as well as from the interest of an attacker in exploiting the threat. There are variants of the STRIDE model which are also popularly used by OEMs and suppliers.

In-Vehicle Infotainment Systems [IVI] and Cybersecurity

With the growing demand for touch-based smart entertainment systems in vehicles, the infotainment system is a very important component which has now become a major differentiator in vehicles. In many countries people spend a major amount of time travelling in vehicles, and an IVI system is popular in reducing the boredom of the journey and providing the driver with an easy interface to access vehicular controls and information with minimal distraction from driving. As the system provides a user-friendly HMI which is also driver-friendly, it is rich in features and a complex system having hardware and software components. It usually provides personalization and customizations to the vehicle user.

The infotainment system hosts the telematics interface of the vehicle and is connected to various other sub-systems of the vehicle. It is a very critical component from the cybersecurity perspective of an automobile. Usually a gateway is used to isolate this system having critical cybersecurity threats from the safety-critical ECUs of the vehicle which could be interconnected over diverse intra-vehicular networks as shown in Fig. 2. Usually all connections to the outside world are channeled through the IVI system. At times firewalls and IDS systems are also used as cybersecurity mechanisms to take care of the threats which are exposed by the IVI system.

Usually the IVI system is feature-rich and rich in the sense that it provides a lot of attack surfaces in a vehicle. Some of the attack surfaces are as follows:

1. USB ports—Usually a mobile device or external pen drives are connected and can easily compromise the security of the system or introduce malware. There have been cases where the USB has been used to attack the IVI system by updating the firmware of the system with a malicious code.
2. Diagnostic port—Allows access to outsiders during maintenance and servicing of the vehicle. An attacker can use this port to carry out an attack on integrity and confidentiality.
3. Multimedia playback systems like software radio and audio players can be used to carry out certain types of attacks.
4. Short-range communication—like Bluetooth or Wi-Fi can be used to carry out attacks related with confidential system information or malware injection.

5. Long-range communication—for data connectivity through a mobile communication channel; usually connects to the Internet, from which remote hackers can carry out various types of attacks. GPS connectivity can also be used to hack into an IVI system.

Fortunately techniques used for securing the IVI system in a vehicle are similar to cybersecurity of computing systems and mobile devices, and lots of best practices, tools, and products are available. And intrusion detection systems, firewalls, gateways and other cryptographic tools, and secure protocols are used to take care of the cybersecurity threats posed by the IVI.

AUTOSAR and Automotive Cybersecurity

Automotive Open Software Architecture [AUTOSAR] is an international standard for development of the software stack for an automotive ECU. The objective of this standard is to ensure interoperability among different ECU suppliers supplying the ECU hardware which has tied up basic software stacks and other system and application software components to OEMs. One more major objective of this standard is to reduce the complexity by standardization so that the suppliers can focus on building more complex application over the standardized stack.

Throughout the development of the AUTOSAR standard, the focus has been on defining the interfaces and verification of the conformance of a developed system, with respect to the standard, by performing conformance testing. As this standard development began almost two decades back, there have been various instances where security considerations have been discussed and have been incorporated into the standard in its various releases. The security focus of AUTOSAR has been to define procedures and interfaces for secure onboard communication. The security-related AUTOSAR [4] components of the basic software stack are distributed as follows:

1. Cryptographic Service Manager [CSM] which is a component of the system services layer
2. Cryptographic Abstraction Library [CAL] which is also a component of the system service layer
3. Secure On-board Communication [Sec-Oc] module which is part of the communication services

These modules are responsible to implement secure communication services by providing the following functionalities:

1. Standard interfaces for providing cryptographic services
2. Cryptographic functionality for which support is provided for Hardware Security Module [HSM] or Software Library
3. APIs for carrying out secure on-board communication through the multiple buses supported by the ECU

ISO 26262 and Automotive Cybersecurity

The ISO 26262 standard [5] deals with the functional safety of automotive systems. A vehicle, overall, is a safety-critical system as a malfunction can lead to losses of lives. All the safety-critical sub-systems in an automobile inherently need to be secure. A compromise on the security of the sub-system can easily compromise the safety of the operation of the system. Functional safety and reliability for an automobile system are well-understood domains having standards, tools and techniques. The overlap between cybersecurity and functional safety is an area that is of interest and further research.

Are security countermeasures sufficient to ensure the safety of an automotive system? To answer such questions; people, tools and techniques from different domains need to come together. Some questions that are relevant in this regard are as follows:

1. Is it necessary to design the system having distinct safety and security goals and approaches?
2. Should safety + security be considered together in the design of the system?
3. At a component level, can a component which is not having critical safety level be compromised by security and can be used to compromise the safety of the system as a via media?
4. Can a security attack lead to a safety hazard?

As illustrated in Fig. 3 all safety-critical systems need to be secure. In case of a cyberattack, the hacker can intentionally cause the failure of the safety-critical component. There could be systems which need to be secure and which are not safety-critical. We could also have systems that are non-safety-critical and not secure in an automotive system.

ISO 26262 [5] defines the safety levels for vehicular components and provides practices to design a safety-critical system in the presence of hazards, whereas ISO/SAE 21434 [7] defines the risk associated with security threats to the system. Is there a need to combine these parameters and together come up with a safety + security number? Ahmad [6] has given a detailed analysis of how security and safety are dealt with in standards like the AUTOSAR. The challenge of managing security is the unpredictable nature of cyberattacks. Whereas in safety we deal with hazards, the cybersecurity attack is active and adaptive as we are dealing with an intelligent adversary.

Fig. 3 Safety and security

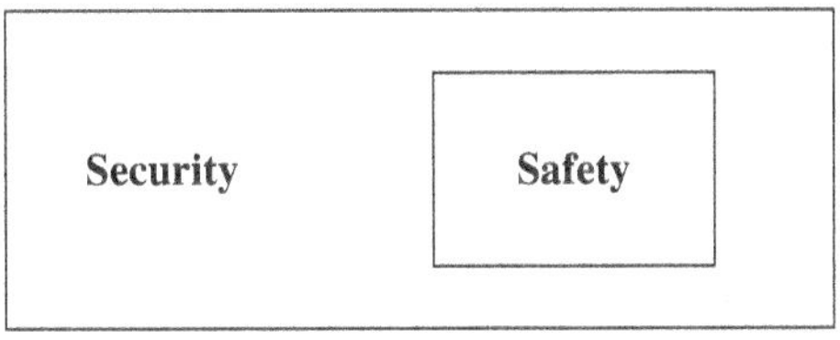

ISO/SAE 21434 Automotive Cybersecurity Standard

The need to have an automotive standard for cybersecurity stems from the fact that multiple parties are involved in the manufacture of a vehicle. The OEM has numerous tier 1 and in turn tier 2 suppliers who supply parts including mechanical, hardware, and software parts that go in a particular vehicle make. As automotive cybersecurity involves security across these components assembled and working together as a system, it is necessary to have seamless implementation of security across these sub systems. This calls for a standardized approach to cybersecurity.

The standards are usually enforced by the OEMs on their suppliers and ensure that the final integrated vehicular system satisfies commonly defined goals on cybersecurity. Standards like the joint ISO/SAE 21434 [7] play a very important goal in fulfilling the following cybersecurity requirements:

1. Clear understanding of automotive cybersecurity terminology across all parties involved in the system development
2. Defining common cybersecurity goals that the system should meet which in turn leads to security goals for the sub-systems designed, implemented, and integrated by diverse parties
3. Creating cybersecurity requirements from the goals for the specific parts of the system made by a supplier
4. Having a trail of documentation related to design, implementation, and testing that can be traced for specific security goals and requirements
5. Satisfying security requirements across sub-system boundaries and software interfaces
6. Verification and validation of cybersecurity at the sub-system and at the system levels
7. Ensuring conformance and auditing that the delivered part satisfies the required level of cybersecurity

Initially SAE came up with J3061 [8] guidelines for cybersecurity. Then later ISO and SAE started work on jointly developing the standard ISO/SAE 21434. Various automotive OEMs, ECU suppliers, chip designers, and cybersecurity companies have come together for the development of this standard which is about to be released. This standard will be applicable for cybersecurity for all road vehicles. The standard defines the activities to be performed and processes to be followed for all the phases of the vehicular lifecycle. Typical vehicular lifecycle stages and the impact of security in these phases are as follows:

1. Research and development—It is good to identify the security threats and their implications to the proposed application.
2. Design and engineering— Security requirements influence the design and implementation of the hardware and the software components of the system. Security requirements, design principles, and test cases are to be used in this phase.
3. Production—The engineered security components are preserved and replicated without compromise of the security.

4. Operation by customer—At this phase the system is open for different threats and attacks. The security provided is able to successfully combat the attacks.
5. Maintenance and service—is when the security-related updates and fixes can be applied to ensure continued security of the system.
6. Decommissioning—is when privacy-related threats are high and security credentials can be misused or stolen for attacks on other systems.

The ISO/SAE 21434 standard does not get into the following requirements:

1. Trying to enforce or recommend any specific security solution, technology, or product
2. Remedial solutions and techniques
3. Telecommunication services, systems, or products
4. Servers, back-office techniques, and their connections
5. Electric vehicle charging technologies and services
6. Requirements that are specific to autonomous and driverless vehicles

The standard deals with three different structured layers. The top layer consists of cybersecurity management across various phases. Next is the risk management across the different phases, and finally there are supporting cybersecurity processes. The approach taken is a security risk-based methodology and risk assessment and prioritization approach.

The concept phase for J3061 begins with "Threat Assessment and Risk Analysis" [TARA] [9]. TARA is crucial for recognition of risks early in the development phase and prioritizing the risk levels. The three fundamental security parameters of confidentiality, integrity, and availability are used for assessing the threats and analyzing those. E-Safety Vehicle Intrusion Protected Applications [EVITA] [10] provide the mechanism for risk assessment at the system level. Threat and operational analysis [THROP] uses a threat- and hazard-based analysis approach for risk assessment. Threat Vulnerability and Risk Analysis [TVRA] and Operational Critical Threat, Asset, and Vulnerability Evaluation [OCTAVE] are some other risk assessment techniques that can be used for security assessment.

Blockchain Technologies for Automotive Cybersecurity

Blockchain technologies for automotive [11] domain are one of the most recent trends getting popular. Blockchain offers a secure distributed ledger and is suitable for taking care of secure storage and integrity of information in the automotive domain. Right from OEM and supplier-specific information to information related to spare parts, insurance, vehicle history, ownership, maintenance repair, etc., blockchains can play a major role in securing applications, since they are dealing with distributed information. It is anticipated that in the coming days the usage of blockchains in the auto industry is going to increase many folds.

Some of the benefits of using blockchain in the auto industry [11] are as follows:

1. Tamper-proof data—Which is reliable.
2. No single point of failure in the system—As shared by multiple parties.
3. Transparency of the data—Everyone can see the entity and the changes made.
4. Identity management—Transactions made by an entity are tied to the identity.
5. Irreversible—Changes made cannot be reversed surreptitiously.
6. Information security—Confidentiality, integrity, and availability.

Automotive blockchain consortium Mobility Open Blockchain Initiative [MOBI] [12] is working on an initiative to standardize the automotive blockchain for vehicular identification [VID]. The objective of VID is to be able to identify a vehicle uniquely across OEMs and make and use the unique ID for other applications for tracking the vehicle like ownership changes, repairs, warranty, etc. The standardization effort also aims to improve vehicular on-road safety and emission norms and looks at techniques for reducing on-road congestions.

References

1. S. Abinesh, M. Kathiresh, R. Neelavenik, Analysis of multicore architecture for automotive applications, in *Conference on Embedded Systems (ICES)*, (Coimbatore, India, 2014), pp. 76–79
2. M. Kathiresh, R. Neelaveni, S. Vismitha, An integrated approach for data security in vehicle diagnostics over internet protocol and software update over the air, in *Computers and Electrical Engineering*, vol. 71, (Elsevier, Amsterdam, 2018), pp. 578–593
3. A. Shostack, *Threat Modeling: Designing for Security* (Wiley, Hoboken, 2014). isbn:978-1-118-80999-0
4. AUTOSAR CP. Specification of Secure Onboard Communication. AUTOSAR CP Release 4.3.1, Document id: 654.
5. ISO26262 Standard. Road vehicles – Functional safety (2018)
6. A.M.K. Nasser Securing Safety Critical Automotive Systems, PhD thesis, University of Michigan, 2019
7. C. Schmittner, G. Griessnig, Z. Ma, Status of the development of ISO/SAE 21434, in *25th European Conference, EuroSPI*, (Springer, Bilbao, Spain, 2018), pp. 504–513
8. M. Steger, M. Karner, J. Hillebrand, W. Rom, K. Römer, A security metric for structured security analysis of cyber-physical systems supporting SAE J3061, in *IEEE International Workshop on Modelling, Analysis, and Control of Complex CPS (CPS Data)*, (2016), pp. 1–6
9. ISO-SAE 21434. Road Vehicles – Cybersecurity Engineering: General Overview. https://www.iso.org/standard/70918.html. Accessed 30 Apr 2020
10. O. Henniger, A. Ruddle, H. Seudié, B. Weyl, M. Wolf, T. Wollinger, Securing Vehicular On-Board IT Systems: The EVITA Project. https://evita-project.org/Publications/HRSW09.pdf. Accessed 30 Apr 2020
11. P. Fragma-Lamas, T.M. Fernandez-Carames, A review on blockchain technologies for an advanced and cyber-resilient automotive industry, in *IEEE Access, Special Section on Advanced Software and Data Engineering for Secure Societies*, vol. 7, (2019)
12. M.M. Castorillo, The World's Largest Automakers, Along with MOBI, Announce a Joint Proof of Concept for the First Vehicle Identity on Blockchain. (2019). https://dlt.mobi/. Accessed 30 Apr 2020.

Autonomous Vehicles: Present Technological Traits and Scope for Future Innovation

Arun S. Tigadi, Nishita Changappa, Shivansh Singhal, and Shrirang Kulkarni

Introduction

There are many major milestones in the rapidly evolving story of self-driving cars. A driverless future seems just around the corner, but it is been a long arduous journey to arrive at this point. Ever since the automobile industry has been in existence, self-driving cars have been a dream. Francis Udina demonstrated a remote control car in 1925 that drove through the streets of Manhattan frightening passersby with an empty driver's seat; it was called the American wonder.

The self-driving car dream was alive and well in 1956, as GM or at least a promotional video leaning self-driving cars would arrive by 1976. In fact, more than 94% of all driving accidents today are caused by some kind of human error. What if automation could help us minimize or completely eliminate driving deaths by taking human error out of the picture? There are also many other profound benefits that could arise from a driverless future. There are also many other profound benefits that could arise from a driverless car in future, such as rather than just focusing on lanes, turns, and traffic while seated in a driverless car, you can even reply to work emails, eat, or relax and make the morning so much more productive. A robotic van that could drive fully autonomously without traffic was developed by Ernst Dickens and his team from the University of Munich in 1986, and by 1987, it drove at speeds up to 60 km/h. His team developed an autonomous Daimler Benz using psychotic computer vision that is focusing on points of interest in the environment and contributed significantly to the Eureka Prometheus project, in the early 1990s. At the

A. S. Tigadi (✉) · N. Changappa · S. Kulkarni
Department of Electronics and Communication, K.L.E Dr. M.S. Sheshgiri College of Engineering and Technology, Belagavi, India

S. Singhal
Department of Mechanical Engineering, K.L.E Dr. M.S. Sheshgiri College of Engineering and Technology, Belagavi, India

M. Kathiresh, R. Neelaveni (eds.), *Automotive Embedded Systems*, EAI/Springer Innovations in Communication and Computing, https://doi.org/10.1007/978-3-030-59897-6_7

same time, Carnegie Mellon University's Navlab in the USA was busy building a steady series of prototypes with ever-improving capabilities.

Their first autonomous vehicle, Navlab1, developed in 1986, managed 30 km/h on the road using multiple sun workstations.

Autonomous vehicles (AVs) are going to influence the near future, yet deployment of this technology continues to be delayed because of the current challenges. There are six separate levels of automation as prescribed by SAE International (2016). These levels range from zero (complete human interaction) to five (fully autonomous) [1].

Independent vehicles can possibly present new and genuine accident dangers, for example, crashes coming about because of digital assaults. There stand a few difficulties in the method for dealing with these dangers and amplifying the advantages of AVs [2].

The advancement in technology is accelerating the growth of autonomous driving. Autonomous vehicles (AVs) are the latest trend that the car industry is moving toward, as AVs will help in replacing humans as drivers who are prone to making mistakes and errors while driving the vehicle. Thus, having AVs on roads will reduce the number of accidents caused due to the human error and hence serve as a means to improve the road traffic safety. However, AVs face innate safety and security challenges that must be addressed before they are deployed for use.

The complex hardware design required to accommodate autonomous systems is shrinking, whereas we can see a sudden increase in the processing speeds. Advanced driver assistance systems (ADAS) are readily used in buses that have amazing features such as pedestrian detection system, adaptive cruise control, collision avoidance, correction of lane, and automated parking facility that can be used in the design of AVs as well [4]. Every novel feature that you want to integrate into AVs such as smartphone integration, entry without keys, and detection of blind spots in the next-generation vehicles brings new vulnerabilities and challenges to the electronic control unit (ECU) [10].

Even with all the research going on, to make the driverless cars a success, the autonomous industry is facing its own set of challenges with respect to the technical implementation of driverless cars such as decision-making, actuation, sensor management, object detection efficiency, safety and reliability, quality of software, computational resources, security and hacking, and privacy, being the essential ones. There is still a long way to go for the industry till it actually reaches its ultimate destination. So this chapter dives into the current state of autonomous driving, the various stages, and the key technological challenges that the vehicle manufacturers must address to get in the game.

Research Motivation and Purpose

Autonomous intelligent vehicles are conventional innovation sets to strengthen vehicle self-sufficient driving totally or incompletely for self-sufficient and security purposes. Generally, self-ruling clever vehicles plunk down with numerous versatile robot advances. In principle, we consider self-ruling astute vehicles as versatile robot stages in this book. Subsequently, condition recognition and displaying, limitation and guide building, way arranging and dynamic, and movement control are the four key innovations present in a smart vehicle, shown below (Fig. 1).

Increasing traffic congestion and accidents has been a major concern associated with the rapid growth in automotive production. Governments across the globe have been improving traffic infrastructure, imposing traffic regulations, and educating people about the same to solve these problems by increasing funds. Furthermore, various research and development projects related to driver assistance and safety warning systems have been launched by research institutes. Therefore, various research works in the domain of intelligent vehicles all over the world have led to intelligent transportation systems (ITS) in improving road safety and reducing traffic accidents since the last decade [3].

Autonomous intelligent vehicles are now widely applied to driver assistance and safety warning systems (DASWS), such as forward collision warning, adaptive cruise control, and lane departure warning. With the development and advancement of the economy and society, the issues regarding traffic safety, energy deficit, and environmental pollution have become grave problems in recent years. Hence, higher volumes of research and applications are inspired due to these problems. In the end, improved traffic capacity and traffic safety can be implemented using computer control, A.I., and communication technologies [4].

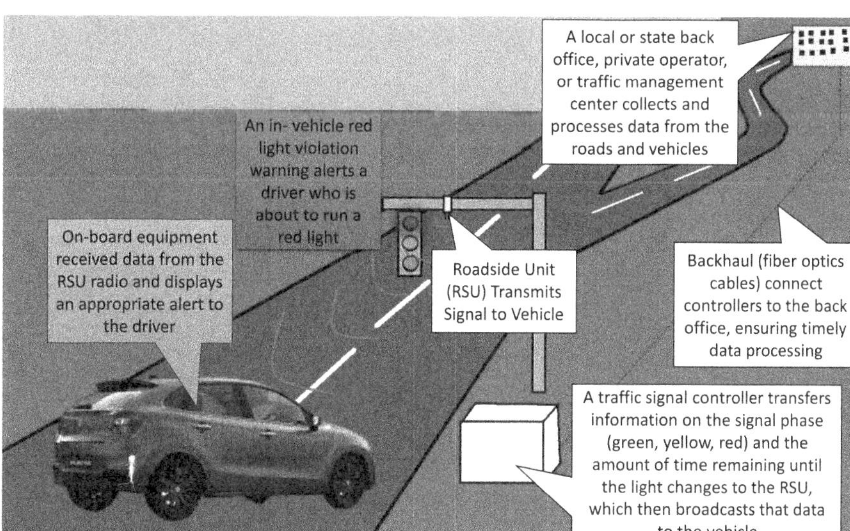

Fig. 1 Vehicle to infrastructure connectivity

Levels of Automation

For the purpose of our classification, we will use the decomposition suggested by the Society of Automotive Engineers in 2014, but the standards are still continuously evolving. The whole autonomy of driving is classified into different levels from level 0 to level 5.

Let us start with full human perception, planning, and control, calling this as level 0. If an autonomous system assists the driver by performing either lateral or longitudinal control tasks, we are in level 1 autonomy. Both the control tasks and the lateral and longitudinal control are performed in level 2 automation, but only in specific driving scenarios. In addition to the control tasks, the system can perform object and event detection in response to a certain degree, in level 3 automation. In the next level, we arrive at highly automated vehicles, where the system is capable of reaching a minimum risk condition, in case the driver does not intervene in time for an emergency. However, level 4 still permits self-driving systems with a limited (operational design domain) ODD. Finally, in level 5 the system is fully autonomous, and its ODD is unlimited, i.e., it can operate under any necessary conditions. If level 0 is the car you drive, then level 5 is the car that drives you as shown in Fig. 2.

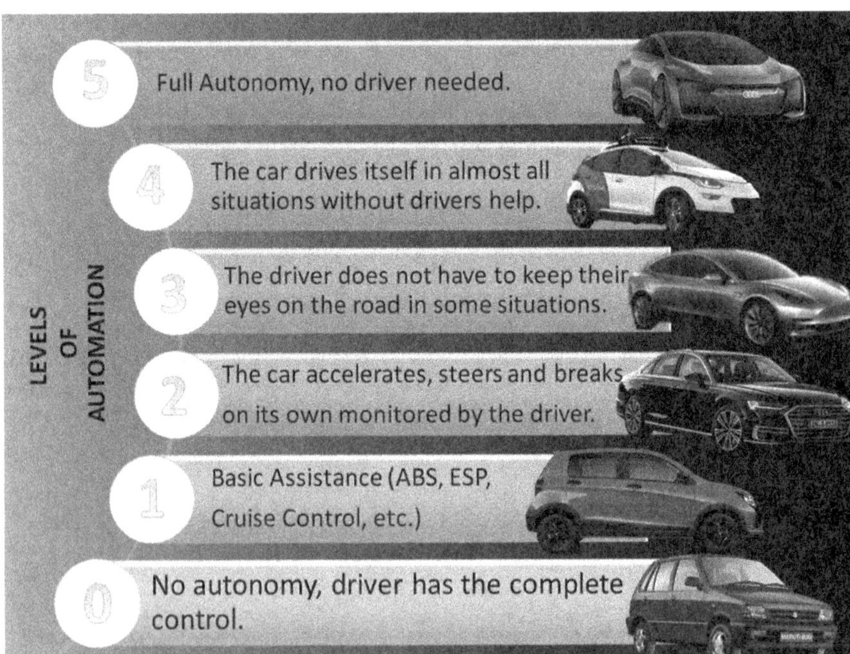

Fig. 2 Levels of autonomy in vehicles

Key Technologies Employed for the Driverless Cars

The design and functioning of passenger vehicles have remained superficially static ever since they have been in mass production, i.e., almost 80 years. They consist of one petrol or diesel engine, four wheels, and the familiar user interface of the steering wheel, throttle, gearshift, and brake. However, the underlying control systems have changed dramatically, in the past two decades. Today's automobile contains a myriad of computers and is no more a merely mechanical device [4].

AVs are foreseen to be driverless. To achieve the perfect driving results, automakers around the world have begun to work to understand the potential and fathom the difficulties that are faced at present. To modify and guzzle existing innovation in traditional vehicles and make an interpretation of them to a close to expected self-governing vehicle would be the main test [3]. Figure 3 depicts probable types of sensors present inside a AV.

The most advanced or best of these cars incorporate some or all of the features listed below (https://www.tempoautomation.com/blog/features-of-todays-best-autonomous-cars/)

- *Lane control*: The capacity to remain securely inside the path is called lane control. It is accomplished by observing separations to path markers, street edges, and adjoining vehicles. A few frameworks utilize the GPS to pinpoint the area of the vehicle.
- *Adaptive cruise control (ACC)*: The already existing cruise control can maintain the constant speed, whereas in ACC it automatically maintains both speed and safe distance from the other vehicles.

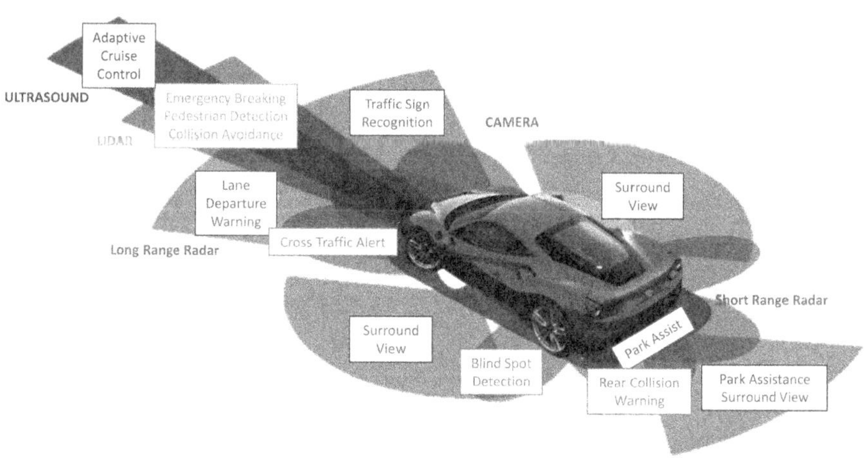

Fig. 3 Different sensors present and the approximate range as different colors

- *Automatic emergency braking system (AEBS)*: It automatically applies braking systems to avoid a collision and therefore is an essential safety feature for completely autonomous vehicles.
- *Light detection and ranging (LIDAR)*: Distance determination and object identification is achieved using this technology. Drones, as well as road vehicles, use standalone mountable units of these.
- *Street sign recognition*: It is a feature that identifies road signs by processing sensor data using a complex software program. Even though workable technology already exists, it cannot be relied upon, and hence this technology will be the subject of research and development for some time to come.
- *Vehicle-to-vehicle (V2V) communication*: Here the multiple vehicles work together and talk with each other to improve the safety of the whole self-driving ecosystem, and this is considered as the heart of a connected vehicle technology.
- *Object or collision avoidance system (CAS)*: This system integrates multiple features, like object detection or identification and advanced emergency braking system (AEBS), to avoid a collision.

Various Technical Challenges in an Autonomous Vehicle

Decision-Making

Recognition is the initial phase in playing out a driving undertaking. The first step towards driving any vehicle is the basic understanding of driving and getting a judgement of the obstacles present on the road. So its basically the observations obtained from the surrounding environment of the vehicle (i.e. road and obstacles...) and making a judgement to drive. This in technical terms is known as decision making and then implementing those decisions into action. To make these decisions, the AV's are programmed with certain plans categorized based on the timeslots that discussed along with some examples below.

In decision-making there are three kinds. They are as follows:

- The first one is long-term decision: how to navigate from Belagavi to Bengaluru or from my home to the workplace, etc. By responding to this inquiry, we have an elevated-level arrangement for the whole driving undertaking.
- The second one is short-term driving: making decisions of changing in lanes or taking a left or right turn at the intersection.
- The third one is immediate decision: making decisions which include control and path planning such as following a lane on a curved road, steering control, acceleration or brake control, etc.

Let us consider an example of proceeding toward an intersection as shown in Fig. 4, including all decisions that are involved in it.

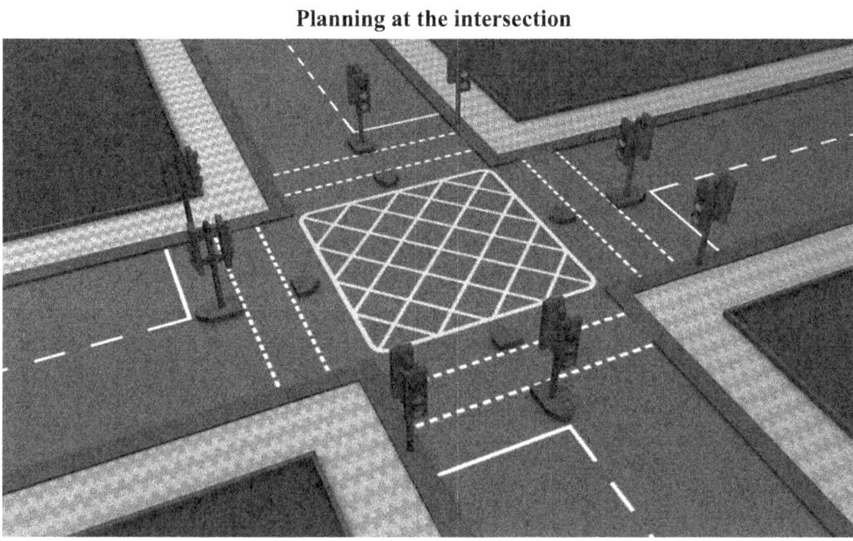

Fig. 4 At an intersection situation for an autonomous vehicle

The drawn-out arranging stage tells you to turn left at this crossing point. Presently, we should investigate the middle of the road and transient choices that should be made. To begin with, let us accept that the crossing point is controlled. That is, there is a traffic light. Since you are turning left, you have to see whether you have to make a path change into a left turning path. At that point, as you are continuing this crossing point, you decide to back off, and it must be done easily with the goal that the travelers do not encounter uneasiness. You have to grind to a halt not long before the stop line, i.e., before a person on foot crossing. These choices made on path changing and halting areas are on the whole transient arranging choices. You additionally need to think and react to conditions that emerge en route.

Despite everything there is still need object and event-detection and accurate response. Imagine a scenario that a vehicle comes all of a sudden infront of you near the turning . So now, now you must decide whether the car has to stop before it can make way for the other vehicle and let it pass or not. Imagine the scenario that a vehicle maneuvers into the turn path right infront of you all of a sudden, out of nowhere. In view of the quantity of potential situations that happen, how does the choice to execute a left turn change if there should arise an occurrence of ordinary driving? These choices fall into the prompt choice class, and they require safe responses from the arranging framework.

Thus, it results in an infinite number of possible decisions that have to be evaluated on different time scales, even if it is a simple scenarios of left turn. This brings up multiple cases that are different for the same intersection crossing scenarios. Every scenario needs a set of choices that are to be evaluated in real-time and update them when new information about the scene is made available. Changing a lane affects various scenarios like where to drive, which cars to pass by and so on. This simple lane change involves atleast three to four decisions to be executed so that the vehicle can be controlled safely. Examples in predictive planning can be as follows: Say a car has not moved for the last 10 seconds and wont be moving for the next few seconds. So is there a way that I can move by it safely, or let me slow down so that the pedestrian can pass-by safely and cross the road infornt of me. So this is the way humans think naturally while driving using their intelligence. Humans make predictions on the next move of the objects and obstacles present on the road before arriving to a decision. This kind of planning relies on preidictions of the actions that other objects in the environment may perform. This increases the complexity in the vehicles when it comes to the tasks that need understanding the environment carefully. But then predictive planning is an important method for AV's because it gives a deep insight into the various scenarios that a vehicle should handle safely without causing aby troubles.

Decision-Making Hierarchy

In this segment we depict the dynamic engineering of a standard self-driving vehicle and remark on the duties of every segment (https://studylib.net/doc/18292501/a--survey-of-motion-planning-and-control-techniques-for-self-driving-urban-vehicles). Driverless vehicles are basically self-sufficient dynamic frameworks that procedure a flood of perceptions from onboard sensors, for example, radars, LIDARs, cameras, GPS/INS units, and odometer. The whole decision-making system scenario of a typical self-driving car is hierarchically decomposed into four important components as shown in Fig. 5.

At the most elevated level, a way is arranged through the road network data. This is trailed by a behavior layer, which settles on a nearby driving assignment that heads the vehicle toward the goal and keeps the standards of the street. This is trailed by motion planning module which chooses a persistent way through nature to accomplish a nearby navigational undertaking. A control framework at that point responsively remedies blames in the execution of the arranged movement.

- *Route Planning*: When driving on a street/road which is slant in nature, the self driving cars' decision making unit should quickly decide the trajectory that it has to use to move through a street from the point where it is currently positioned to its destination and also decide on whether to increase or decrease the speed or apply the brakes (depending on whether the car is moving in the upward or downward direction on the slant street). Next, represent the road network in the form of a directed graph. In this graph the weights represent the cost of passing over a segment of a road and can be used to find a route with a minimum-cost path on a directed graph obtained from the road network [5].
- *Behavioral Decision-Making*: Once the routes for travelling are found, the AVshould select the proper path and follow the selected route for the journey. It should communicate accordingly with other users on the road (other car drivers etc...) for smooth manuever, following all the rules of road safety. It is the responsibility of the behavioral layer for selecting the appropriate driving behavior at any given instance of time in synch with the newly learned behavior of the other

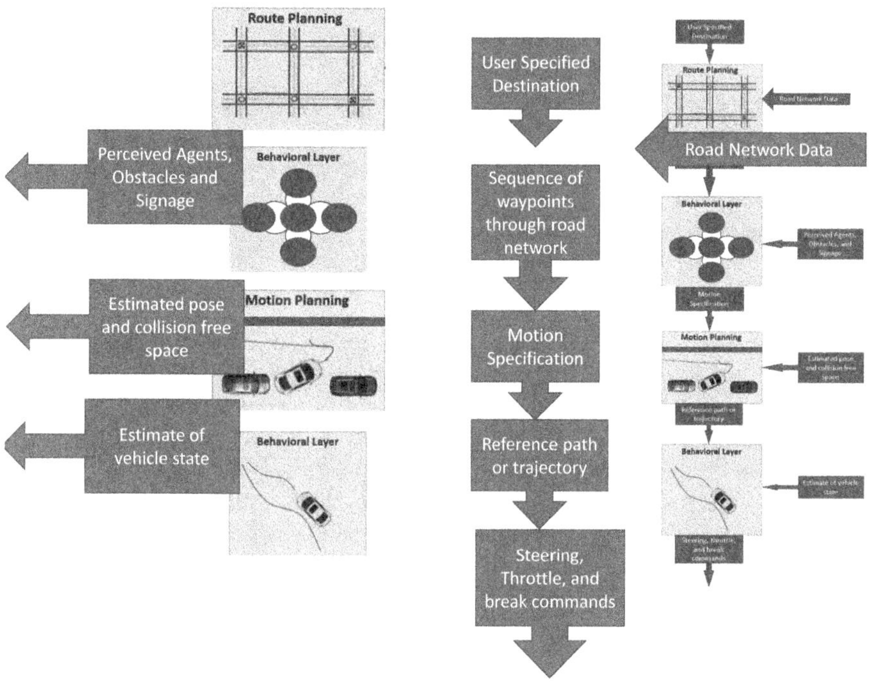

Fig. 5 Illustration of the hierarchy of decision-making process

participants (car, obstacles, humans etc...) on the road, the condition of the roads and also take into account the signals sent by various traffic signals encountered on the selected path [5].

- *Motion Planning*: Once the driving behavior is decided (e.g. taking a U-turn, lane changing or taking a left turn etc) by the behavioral layer, it sends this behavior for conversion in to a path or trajectory that can be stored and used by the low level feedback controller. This path should be practically feasible such that it is convenient for the on-board passengers, avoid any form of wreckage or accidents as the obstacles (detected by on-board sensors) are encountered on the travelling path/road.
- *Vehicle Control*: In the next step, the feedback controller sends the accurate inputs to the actuator to proceed forward with the planned motion and correct the tracking errors in case any encountered. All the tracking errors that occur when the planned motion is in progress are due to the imprecise vehicle model. Hence the closed loop system should be have high durability and stability.

As we know the self driving vehicles are complex systems that are split in to a group of complicated decision making problems, where the the result of the previous stage acts as the input to the next stage and keeps evolving. This capability of decision making has allowed the adoption of well-developed techniques for finding unique solutions from various research areas.

Actuators

Over the years vehicles have evolved from simple electromechanical systems to complex computer-controlled networked electromechanical systems. Apart from solving one of our problems, these changes have led to new problems like accidents leading to death or injury, pollution, traffic issues, fierce competition in terms of cost, and the depletion of fuel reserves.

- Drive by Wire

  Traditionally vehicle functions were achieved by mechanical linkages, but today electrical or electro-mechanical systems are used for performing the same functions. This technology is known as drive-by-wire technology in the automotive industry. This technology uses electromechanical actuators and human–machine interfaces such as pedal and steering feel emulators. Various mechanical components are eliminated from the vehicle.

  Brake-by-wire is a technology within which almost all the mechanical and hydraulic components of traditional braking systems are replaced by electronic sensors and actuators to apply brakes in a moving vehicle. It is a meld of ECS along with a group of electromechanical actuators and a brake pedal. The use of brake-by-wire systems has many benefits like reduction in weight and space, low operating noises and vibrations, and quicker response time because of the absence of mechanical linkages, which might lead to shorter stopping distances.

  Due to the security-critical nature of brake-by-wire systems, incorporating a degree of fault tolerance is important. The foremost commonly suggested solution is a redundant or back-up brake, like a traditional mechanism, which might be brought into activation in the event that the first system suffers a failure (Fig. 6).

  Electronic throttle control (ETC) is an electronic part of the vehicle which helps in interfacing the pedal to the throttle, to form a mechanical linkage. The three parts of any common ETC system are (1) pedal module, (2) throttle valve (controlled by Electronic Throttle body-ETB), and (3) power train. The ECM is a sort of electronic control unit (ECU), which is an embedded structure that uses programming to work out the perfect throttle position by estimations from data evaluated by various sensors, including the pedal position sensors, engine speed sensor, vehicle speed sensor, and journey control switches. The electric motor is then used to open the throttle valve to the vital point.

  Most of the faults that occur in Throttle-by-Wire systems occur due to pedal or throttle position sensors because of wearing out, skipping or emitting erratic signals, throttle body motor failures, and numerous electrical problems, like loose or corroded, wiring connectors are also common (Fig. 7).

- *Steer-by-Wire (SbW)* frameworks are an extra advancement of the present electrical force controlling frameworks. There is no constant mechanical connection between the hand wheel and the street wheels. The guiding order is transmitted absolutely electrically. An actuator on the hand wheel delivers the necessary directing torque (input actuator) to cause a credible guiding to feel for the driver. Another actuator on the controlling apparatus guides the street wheels (directing actuator) (Fig. 8).

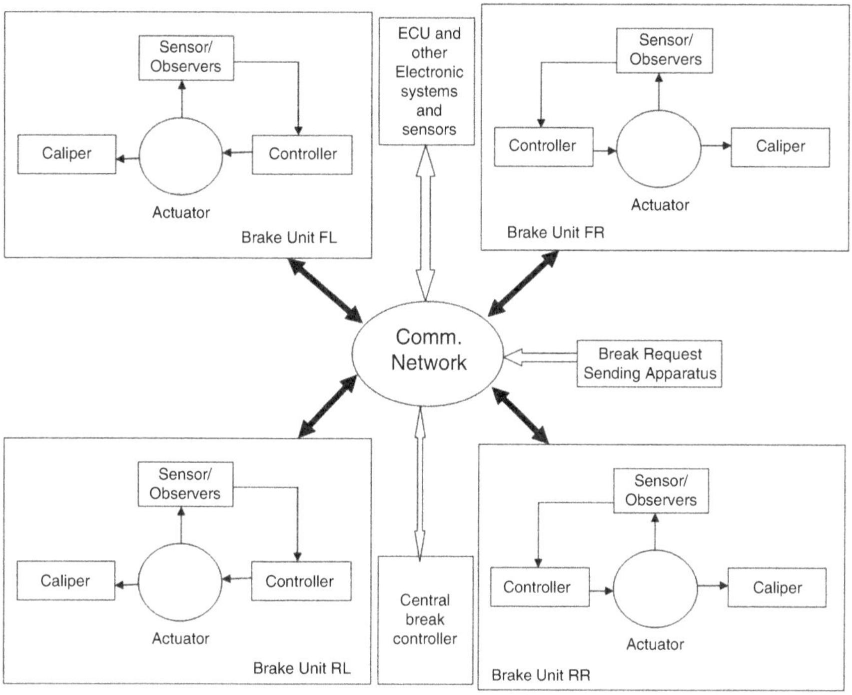

Fig. 6 General structure of a brake-by-wire system

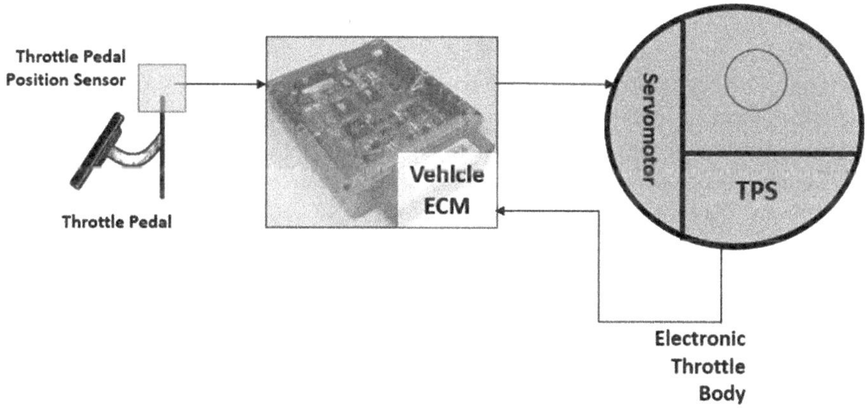

Fig. 7 General Structure of a throttle-by-wire system

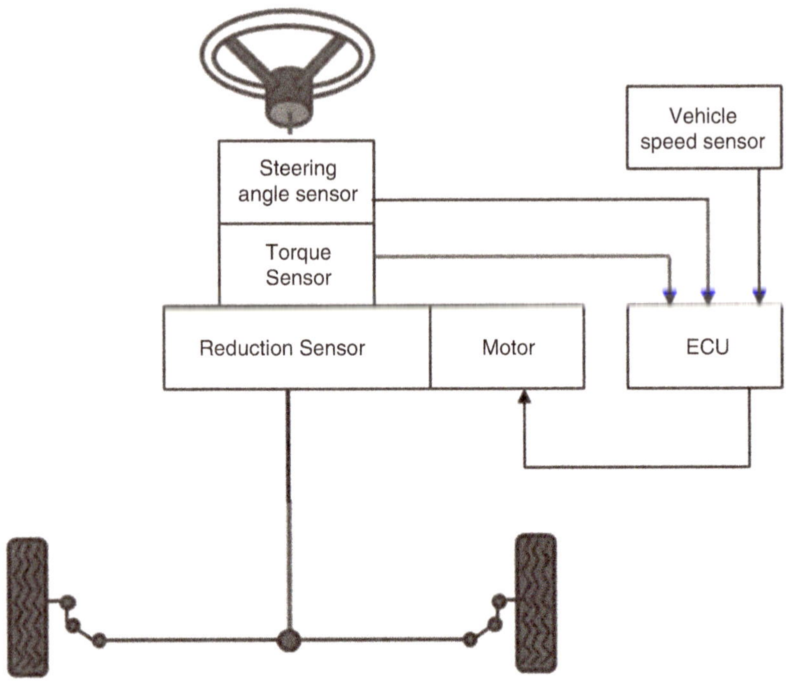

Fig. 8 General structure of steer-by-wire system

Because of the nonattendance of a human driver, supervision, and disappointment, revision must be actualized in electronic frameworks. For this, angles, for example, actuator topology, important level of excess, suspension kinematics, just as structures of controllers must be thought of.

As actuators are totally constrained by electronic frameworks, incitation with regard to mechanized driving is emphatically connected to car by-wire frameworks, for example, steer-by-wire or brake-by-wire. Berg miller broadens these contemplations toward a well-being idea for the trial vehicle MOBILE of TU Braunschweig. The vehicle highlights brake-by-wire, steer-by-wire, and throttle-by-wire at each wheel independently. All things considered, the vehicle is constrained by a human driver.

Horwick records security prerequisites for vehicle activation frameworks. He requests needlessly controllable brakes and repetitive guiding actuators. These are upheld by producing yaw second by differential slowing down if there should be an occurrence of directing misfortune. Moreover, inertial sensors and associated movement estimation should be planned needlessly.

Raste has thought about security objectives and well-being prerequisites for a SAE level 3 framework, too. As one can understand, Raste perceives trajectories as inputs to the actuation systems in the vehicle. For the actuation systems in vehicles, he introduces redundancy structures that have two parallel paths. The normal operation is impelmented by the first path, whereas the second path is a hot standby. Incase a malfunction occurs in the first path, the second path takes over the control. So the elements such as actuators, sensors, reduandant design of controllers and power supply are satisfied. Raste lists out additional safety goals with respect to steering. To sum it up he says unintended steering should be avoided and intended steering should be ensured.

Reschka presents general consultations with respect to useful well-being of mechanized vehicles [6]. Reschka additionally for the most part requires excess plan of sensors, actuators, and force supply for guaranteeing useful security of mechanized vehicles which likewise applies to incitation frameworks.

Sensor Management

There are many shortcomings when it comes about the sensors used. Shortcomings like:

- Sensors are very expensive
- Risks in perception error
- Machine learning becomes unstable under changing conditions
- Vulnerability to attacks.
- Detection range is too short and uncertain.
- High processing power demand

• Sensors are very expensive

In present days, sensors are very expensive, and autonomous cars are occupied with many sensors. The total cost of sensors lies between $100k and $200k for a single vehicle. This amount is not affordable by every common person. Very few people can afford up to $100k for autonomous driving. This price needs to be affordable so that everyone can afford an autonomous driving.

• Risks in perception error

Due to increasing prices of sensors, we use camera-based sensors which are relatively low cost when compared to LiDAR and RADAR systems. But these camera-based sensors have fatal perception error.

By using camera-based sensors, we try to reconstruct a 3D image in terms of 2D. But the reconstruction is not efficient. There are many advances in these area like FPGA-based stereo algorithms, GPU-based parallelized implementation, algorithms such as Semi-Global matching and Structure from motion and its variant, but it is not suitable all the time.

Considering if it is night and there is a person standing, we just can make out a shadow in the camera. So, the camera does not understand if there is a person or just a shadow of a tree or street lamp. This leads to risking of life. A normal camera does not provide sufficient information to make a decision. The information received is not enough for detecting, extracting, and classifying objects.

- Machine learning becomes unstable under changing conditions

Machine learning is a tool that is used to translate the sensor data into perception, prediction, and decision of the trajectory of the car. In self-driving models, there are tens of thousands of neurons and tens of millions of connections. This is a very large model. The complexity of perception is higher than the complexity of prediction, which is higher than the complexity of planning. Normalization adds up to the complexity as it eliminates the variance in sensor data because of different conditions of the environment.

We can assume the total complexity of driving as

$$\text{Complexity}(\text{driving}) = \text{Complexity}(\text{normalization}) * \text{Complexity}(\text{perception}) \\ * \text{Complexity}(\text{prediction}) * \text{Complexity}(\text{planning})$$

If we had to do this improvement without the better sensor data, we need to put ten computers in a car instead of one computer, and this is practically not possible.

- Vulnerability to attacks

Self-driving cars need to cope up not only with just the situations of real traffic but also of the sudden attacks by criminals, terrorists, and other adversaries. We obviously cannot accept our car to freeze when under attack. A problem with cameras is that they can be easily spoofed. Suppose the pranksters draw the realistic image of a tunnel on the house wall, and it may be misunderstood by the controller and the vehicle may run into the wall. But this might happen. LiDAR can also be spoofed, by making the car to see objects while do not exist.

There is no way to provide 100% safety, but we can try to make the work of these culprits exponentially more difficult.

- Detection range is too short and uncertain

Good cameras with large sensors and good optics can resolve to long distance. The problem is that the further away the sensor needs to see, the more pixels it requires. More pixels equal to more computing which further equals to more cost to more hardware. Cameras can be effective only for the use of 3D type of analysis in short distances. LiDAR range is limited by eye safety. The amount of light emitted that can enter an eye is limited by the laser light's average and momentary power depending on the wavelength. Change in weather, such as fog, reduces the range of cameras and LiDAR.

- High processing power demand

The process of data fusion is very hectic for the computer resources as it collects all the sensor data. Even to get the data into a same coordinate system takes a lot of computing power. Data correlation between camera images and LiDAR data can also fail due to various reasons (Fig. 9).

Sensors that are utilized in AVs are camera, RADAR, sonar, LiDAR, a GPS, an inertial measurement unit (IMU), and wheel odometry. Sensors are utilized to gather information that is dissected by the PC in self-governing vehicles. It controls the brake, steer and speed of the vehicle.

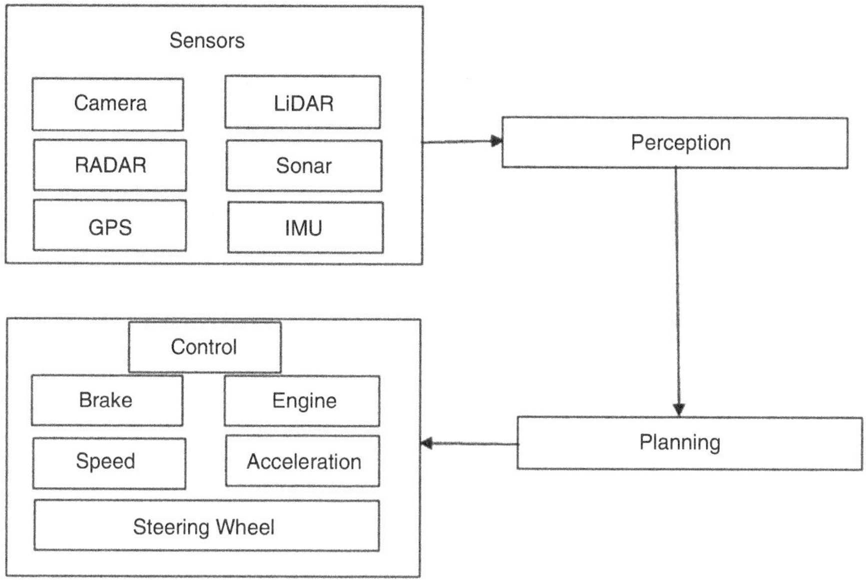

Fig. 9 Block diagram of autonomous vehicle system

- *Camera*: Cameras were the principal sort of sensors to be utilized in the AVs and are currently the primary decision for most controllers used. Camera empowers the vehicle to truly envision its environment. Cameras are profitable at the gathering of surface comprehension; they are extensively open, and more moderate than RADAR or LiDAR. The latest HD cameras can make a large number of pixels for each packaging, with 30–60 edges for each second, to develop continuously incredible imaging. Since the camera is a key sensor in independent driving, the usages of camera in self-driving vehicles are endless. Cameras can be part of human–machine association.
- *RADAR*: RADAR means Radio Detection and Ranging. RADARs are fixed into a vehicle for different purposes such as adaptable journey control and powerless side exhortation crash alert. Relatively few sensors measure speed by calculating the complexity between the two readings; RADAR uses the Doppler Effect for direct measurement of speed. Microwave RADAR works at 77 GHz and can gage vehicles moving around 200 m away. RADAR is more capable than cameras and LiDAR in picked conditions like bad weather conditions.
- *LiDAR*: LiDAR represents Light Detection and Ranging. LiDAR utilizes an infrared laser shaft to appraise the separation between the sensor and a close by object. The current LiDAR's utilization light of frequency is around 900 nm. LiDAR has higher spatial objectives than RADAR, as it has progressively focused laser beam with greater number of clear layers in vertical direction, and the high thickness of LiDAR center per layer.

- *Sensor Fusion*: Sensor blend is the system for the blend data passed on from unique sources with the ultimate objective that the insightful information is made. On the independent vehicle, we have a camera in order to clone a human vision, yet the information of the hindrance partition will be the best expanded through the LiDAR or RADAR. Accordingly, sensor blend of camera with LiDAR or RADAR data is noteworthy, and joining information from LiDAR and RADAR will give dynamically certain information about the division of the deterrent enveloping the vehicle.
- *Sensor Fusion for 3D object detection*: Sensor combination from camera and LiDAR information gives an answer regarding equipment intricacy of the framework; just two sorts of sensors are incorporated, i.e., camera for vision and LiDAR for snag discovery. They supplement one another (Fig. 10).

The advantage for this is having a reasonably straightforward arrangement, which is the present and future pattern in neural system advancement with the motto "little neural nets are wonderful."

Sensor Fusion Object Detection and Tracking

Moving item identification and moving article following is one of the most testing parts of AVs space. Prior methodologies in moving object recognition and following were centered on melding sensor information, which follows following alongside extra data from a simultaneous localization and mapping (SLAM) module. In sensor combination, the information from camera, LiDAR, and RADAR, we have to apply the low-level combination on RADAR and LiDAR information that has been pre-prepared. At that point the cameras are considered as contribution to this melded data which is a piece of a significant-level combination. Restriction and mapping is comprehended to be low-level combination, while the aftereffects of elevated-level combination are recognition and characterization (Fig. 11).

The following framework can be exchanged between a point and a 3D box models [5] dependent on the separation of the article from the vehicle. The future patterns in the data about the urban traffic are being investigated to improve the following framework ability.

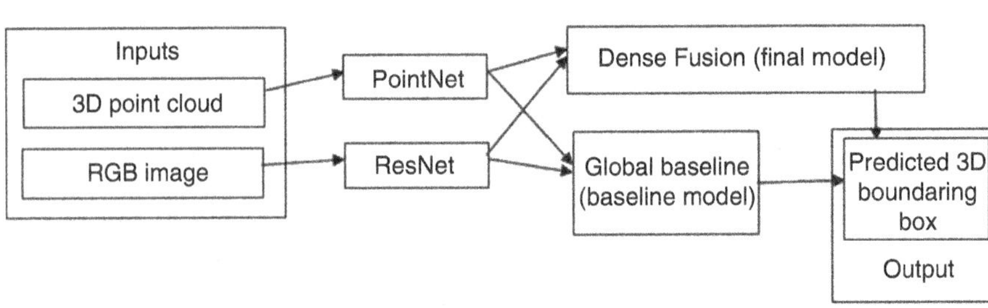

Fig. 10 Block diagram of 3D object detection form image and 3D point cloud data

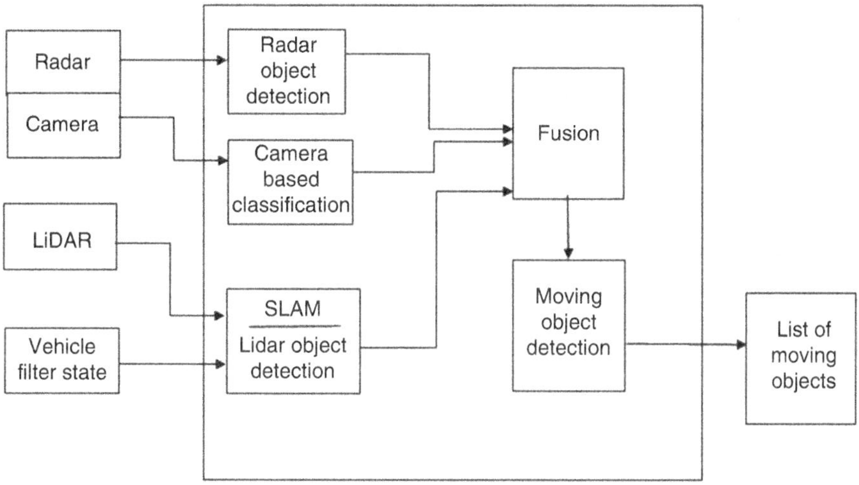

Fig. 11 Multiple sensor perception system

Object Detection

Object detection is a computerized technique that locates instances of objects in images or videos. Object detection algorithms basically include machine learning or deep learning to produce meaningful results. In autonomous driving systems, the object detection is one of the most essential factors in autonomous navigation, as this allows the car controller to account for obstacles when considering possible future trajectories. It follows that we desire object detection algorithms that approximately accurate. Many high-quality object detectors have seen shocking improvements in recent years [7]. However, in many leading techniques the problem of object detection is not mentioned. These problems in autonomous vehicles contain unique challenges. These challenges include the following:

- Car controllers must often solve optimization problems at least once to achieve feasible control; this means that the car controller must receive detection results in a similar time span.
- The entire detection pipeline prevents the use of image preprocessing which often boosts the detection performance, such as extracting SIFT descriptors, region proposals, or sliding windows. Therefore, a detector for an autonomous car must be approximate while operating on strictly the input image.
- Car computing systems are frequently heavily memory constrained, so it is often absurd to store and run detectors with large number of parameters, especially with large input image volumes. In particular, this constrains the depth of neural network approaches.

Different Models Used for Object Detection

- Over-Feat: It is a convolution neural network model that was released in 2013 which jointly performs object recognition, detection, and localization. Over-Feat is one of the most successful detection models till date, winning the localization task in the imagenet large scale visual recognition challenge 2013. Over-Feat has eight layers which depend on an overlapping scheme that produces detection boxes at multiple scales and iteratively assembles them together into high-confidence predictions.
- VGG16: VGG16 was an attempt to usurp Over Feat's dominance in object classification and detection by exploring the effects of extreme layer depth. A 16-layer and 19-layer model was produced, setting new benchmarks on localization and classification in the Image Net ILSVRC 2014 Challenge (Fig. 12).
- Fast R-CNN: Fast R-CNN [8] is a model which by improving the speed it tries to capture the accuracy of deeper models. Fast R-CNN predicts on region proposals, and to speed up the training and prediction time of the model it utilizes the shared computation per region proposal and truncated SVD factorizations (Fig. 13).
- YOLO: YOLO is a model that operates on images directly while treating object detection as regression instead of classification. Thus this model has the lowest performance when compared to others, but it is mainly known for its high speed. As shown in Fig. 14, it predicts astonishingly up to 45 frames/s. YOLO is quite large at 12 layers.

Simple Net—In this work, we investigate the possibility of using simple techniques to construct an object detector that excels in the metrics provided by our autonomous driving environment; that is, we train a detector in such a way that it fits in a confined memory, it anticipates in real-time, and has acceptable precision to be used in an autonomous driving platform.

Safety and Reliability

Aligning safety and security is crucial for autonomous vehicles, since any of failures or attacks may lead to safety losses as seen in Fig. 15.

AV's can have new features added to them regulary, that makes the self driving cars more comfortable, but with every new feature that is introduced there also comes an attached technical challenge, that has to be fixed for smooth functioning. Guidelines in United States are voluntary at federal level, but across the state the laws vary. And there are no specific standard for finding out whether they are properly implemented or not. Human-controlled driving these days is already a remarkably safe hobby—in the USA, for every one hundred million miles drive there is atleast one death. So basically what the car industry is saying is that with self driving cars on roads will help avoid/ prevent the road accidents and deaths due

Fig. 12 16 Layer VGG convolution neural network (CNN)

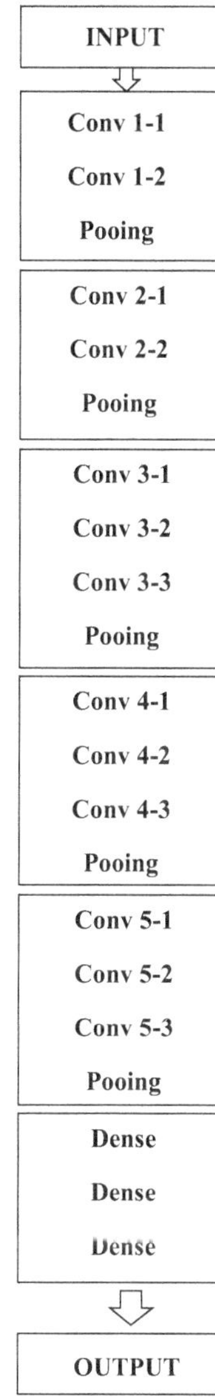

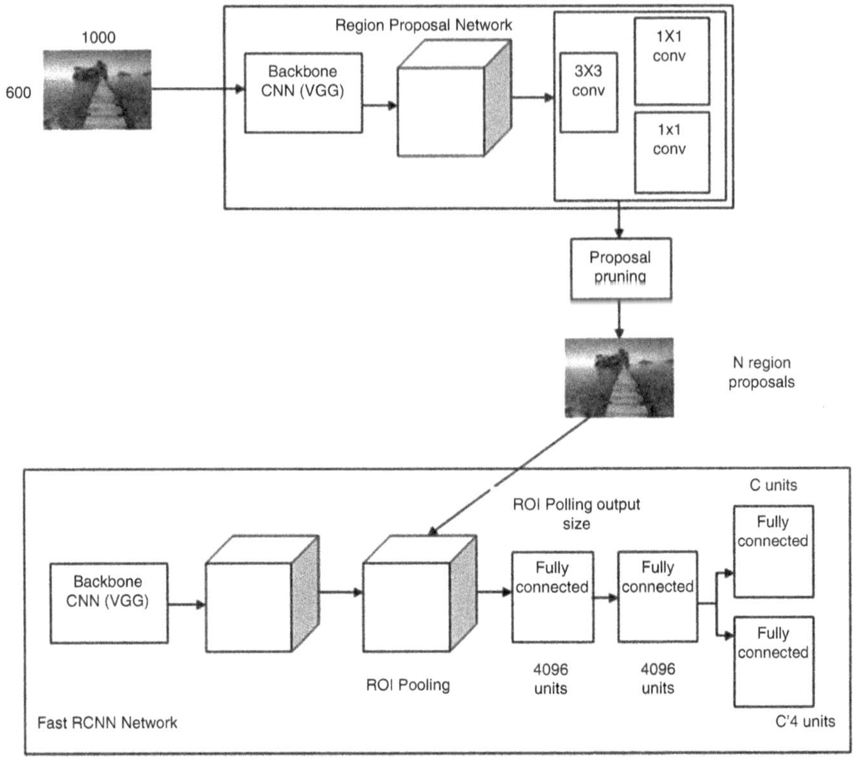

Fig. 13 Block diagram of Fast R-CNN

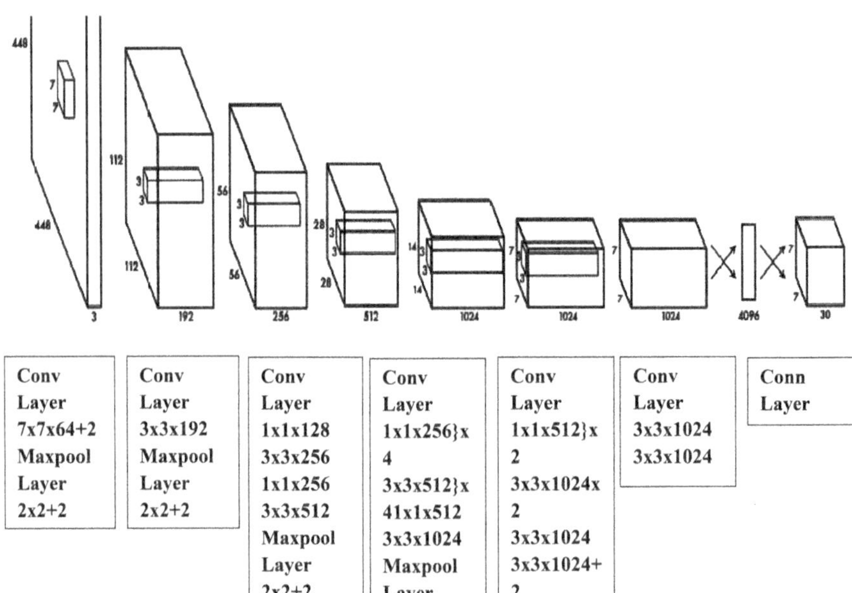

Fig. 14 Different layers for YOLO

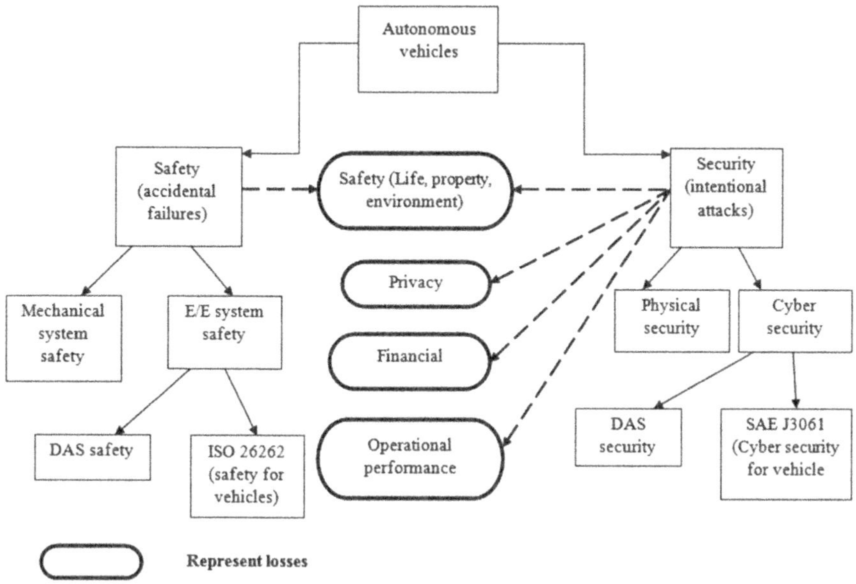

Fig. 15 Safety and security issues in autonomous vehicles

to reckless human driving. Does it have to be only 10% safe or only 100% safe? It does not matter or there is no need to exactly define what percent is actually safe for self driving cars. The thing that has to be focussed right now is that the AV's should make driving an all together a safe, secure and happy experience for its customers and in the mean time try to avoid the accidents.The next big thing that has caught the world's attention is security in automotive computing. It has become a predominat challenge that needs to be addressed quickly. Security has become an important feautre that the self driving cars have to get through. Apart from securing individual cars from attacks the makers should also concentrate on safeguarding the attacks on system level.

Threats Involved with AVs

Threat #1: An Unregulated Industry

Although more than 200 car businesses are jumping into the independent vehicles area, there is not sufficient stable information to make a baseline for protection standards as information approximately the era is confined. As the industry is unregulated, it is brilliant for manufacturers, but terrible for clients.

Prior to buying an AV, the buyer should look up for its VIN number and then find out about it as much as they can before the actual purchase of the self driving car.

Threat #2: More Accidents Blending Self-Driving and Manual Cars

Highways and facet roads around the entire world have not yet been optimized for autonomous cars. Driving is unpredictable, and on account that each possibility needs to be programmed into the automobile, injuries and unexpected effects will appear. The car may not have the proper software program to recognize a way to navigate extreme climatic situations or intricate congestion patterns.

If a challenging situation emerges then the passengers of self driving vehicles should be mentally prepared and ready to drive the car, even though at times self driving cars tend to give the passengers an impression of false sense of security and safety.

Threat #3: Vulnerability to Hacking & distant control

Any laptop tool related to the net is at risk of hacking. These automobiles additionally rely closely on the software that runs their additives, and if a hacker gets into the system, they will manipulate each issue of the automobile.

As we know, softwares are susceptible to hacking, so we should note the other dangers that the AV's software may encounter such as personal data of the people on board being stolen and illegal remote access through the cellphone connected to the car through Bluetooth network. Self-driving automobiles may also be more susceptible to computer viruses.

Threat #4: Computer Malfunctions

AV's are mainly made up of 100's of computers, that means a lot of technology, so anything can go wrong and lead to failure. The software program that runs self-sufficient motors is legitimately sophisticated. Correctly controlling sensors at the rear digicam is moreover a problem. A completely risky glitch is the manner to recognize when to execute an instantaneous forestall while someone is inside the crosswalk ahead of the auto.

The human issue also confounds developers of self-using automobiles, and that they have yet to work out a way to make an automobile decide between two bad options: hitting a pedestrian or another car. Self-using motors are imagined to perform better than humans, but in these styles of conditions, "better" may also be subjective because self-riding cars can also be more risky.

Threat #5: Exposure to Radiation

The regular cars mainly consist of the complex hardware components such as GPS, Wi-Fi, Bluetooth etc. that need the use of electromagnetic (EM) radiation for in-car communication aswell as for out of the car communication. Due to which the drivers are exposed to long term EM radiation. This increase in exposure to EM radiation can cause a lot of serious health concerns such as respiratory problems, headaches due to migraine, vision impairment and eye problems, exhaustion, and lack of proper sleep.

As experts predict, autonomous cars will save lives and will be safer than the cars that have human drivers. The benefits of having self-driving cars is that the accidents will be prevented. But this will take some time. Before that becomes a reality, the car industry should try to resolve the existing concerns and issues by developing appropriate guidelines for safety first.

Typically, in a car accident case, a driver is found guilty if he breaks the road safety rules or has been driving in a negligent manner like drunk driving or texting while driving. But when it comes to AV's which is a computer program, wont be making such errors (like drunk driving etc), This makes it difficult to find out if the program or the driver of the AV was guilty for the accident. For example, the company that manufactured a driverless car X causes an accident, then in this situation the company can be held responsible and the same company's insurance policy can cover the damages caused by the accident. But, the challenge here is about holding the driverless car responsible for such accidents, because it involves paying out the insurance and since there are no proper guidelines about how to handle this situation, hence the law makers have to come up with new policies that can cover these issues over time.

When accidents are caused by AV's, it becomes exceptionally difficult to establish the liability that will involve finding out where and how exactly the error occured. So a software error can hold the AV developer as guilty for the accident. This leads to the next issue i.e; whether the developer's fault is independent from that of the manufacturer's fault. As experts predict, autonomous cars will save lives and will be safer than the cars that have human drivers. This also parades the query of whether an automobile's software program developer's fault is independent from its producer's fault. For example, if Volkswagen outsourced the tech of its self-riding vehicle software, then one among its cars precipitated a coincidence because of a glitch in its programming, it may not be at once apparent whether or not Volkswagen or the company which evolved the program is responsible. This will make the issue of product liability a manner more difficulty inside the realm of vehicle coincidence (Table 1).

Table 1 Various eras of safety in automobile industry

From 1950 to 2000, the features of safety and convenience was developed. This included the cruise control, seat belts, and ant lock brakes	
	From 2000 to 2010, the features of advanced safety were developed. This includes electronic stability control, blind spot detection, forward collision warning, lane departure warning
From 2010 to 2016, the features of advanced driver assistance were developed. This included the rear-view video systems, automatic emergency braking, pedestrian automatic emergency braking, rear automatic emergency braking, rear cross traffic alert, and lane centering assist	
	From 2016 to 2025, the features of partially automated safety are developing. This consists of lane keeping assist, adaptive cruise control, traffic jam assist, self-perk
Beyond 2025, the fully automated safety features will be developed. These might also include the highway autopilot	

Security and Hacking

The potential outcomes of independent vehicles gracing our streets are drawing nearer to reality continuously. Uber and Tesla the famous transport companies are experimenting on cars that make use of computer OS's to give commands to the human drivers. On the other hand Google's driverless car has already ventured on roads few years ago.

Be that as it may, how sheltered and solid are driverless vehicles? Being tech-worked and with both digital and physical segments, any conversation on their security must include both cyber security and furthermore physical street well-being.

Security Threats
How safe and reliable are driverless vehicles is the biggest question. If we are discussing about the security issues and solutions for AV cars, then that discussion should consider both cybersecurity and physical road safety, as these cars are operated by technology that have both the cyber and other physical components. The following are the security threats which we have identified and have a higher degree of influence on autonomous vehicles.

- In future the criminals will hack cars for ransom, and after paying the money, they will allow you to either come out or go inside the car.
- Terrorists may hijack the network and take control over a transport system. Hacking a network of an AV can cause major crashes by disabling many of the important sensors leading to huge confusion.
- Hacking the car's OS and destroy its contents could harm the user both financially and physically.
- Hacking into an autonomous car could lead to some more serious issues such as expose of your personal data including your travel plans to an unknown person. With this information, anyone could potentially track the user with an intention to rob or assault or even can keep him/her as hostage. The hacker may also redirect the passenger to an unknown and isolated location.

As the technology evolves, AVs will be able to control any of the smart devices such as the TV, heater, door, or gate. Hackers could use these features for their access to the house of the vehicle owner (Fig. 16).

Hackers and Types
There are generally six types of Hackers:

- *White Hat Hackers*: Approved or the guaranteed programmers who work for the administration and associations by performing infiltration testing and recognizing escape clauses in their cyber security. They likewise guarantee the insurance from the pernicious digital dangers. They work under the standards and guidelines gave by the administration and express, that is the reason they are called Ethical programmers.

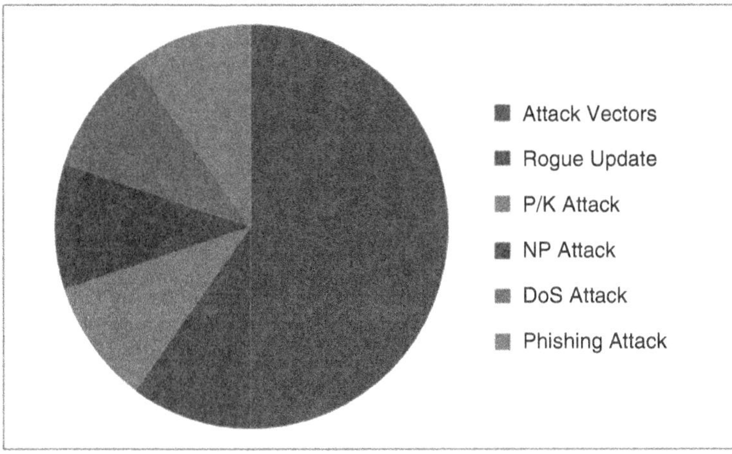

Fig. 16 Cyber Vulnerabilities in an AV

- *Black Hat Hackers*: They can pick up the unapproved access of your framework and crush your fundamental data. They are viewed as hoodlums and can be effortlessly distinguished on account of their vindictive exercises.
- *Gray Hat Hackers*: They fall some place in the classification between white hat and black hat programmers. They are not legitimately approved sort of programmers as they work with both great and awful goals.
- *Script Kiddies*: They are the most riskiest individuals as far as programmer's network. A Script kiddie is a person who doesn't possess the high level skills of a hacker, but uses tools and scripts by downloading them internet. These are normally provided by other hackers for hacking online. Script kiddie uses these for personal gains and his hidden agenda.
- *Red Hat Hackers*: They are additionally called as the hawk looked at programmers. Like white hat programmers, red hat programmers likewise plan to end the dark hat programmers and forestall any conceivable risk.

V2X and Security Challenges

As we edge nearer to the wide execution of wise transportation frameworks, the need to expand the limits of sensor-prepared vehicles past the individual vehicle has gotten more articulated than any other time in recent memory. Research and normalization endeavor toward vehicle-to-everything (V2X); innovation are expected to empower the correspondence of individual vehicles with each other and supporting street foundation all the while. The subject has drawn enthusiasm from countless partners, extending from legislative specialists to car makers and different portable system administrators. With intrigue sourced from numerous unique gatherings and an abundance of research on an enormous number of points that are being done in this field, attempting to get a handle on the master plan of V2X advancement is an overwhelming assignment.

The main intent of V2X is to increase the safety of the road and to improve traffic management. V2X correspondences allude to data trade between a vehicle and different components of the ITS, including different vehicles, pedestrians, and transport framework (for example, traffic lights and signs). The recent years saw quick development of extension in the metropolitan regions alongside increment in the vehicular traffic to and from huge urban areas. Along these lines, traffic blockages and street mishaps are on the ascent as of late (Fig. 17).

Shrewd transportation framework applies ideas of information handling, correspondence, and different sensor advances to vehicles, foundation units, and side of the road clients to build security and efficiency of the general transportation framework. The heterogeneous network comprises of two main sub-networks as shown in Fig. 18.

Vulnerabilities within vehicular communications lead to four vehicular cyber security challenges described as follows:

- *Limited connectivity*: In spite of the fact that the outside network of vehicles is expanding, most vehicles do not be able to refresh their product through Over-the-Air (OTA) refreshes, which would push vehicles to empower different vehicles to consistently be secured against the most recent digital assaults [9].
- *Computational performance*: Vehicular computational execution is commonly constrained, when contrasted with the computational presentation of a gadget, for example, a PC. It is constrained because vehicles have to withstand high temperatures and vibrations as compared to an average laptop or a table top PC, as vehicles lifetime is longer when compared to a table top PC or a laptop. Because of their computational weakness, vehicles are more inclined to be hacked than PCs. The restricted computational presentation of vehicles will likewise imply that some vehicular cyber security arrangements will have too high an overhead to be actualized to forestall such assaults.
- *Unpredictable attack scenarios*: A vehicular engineering can be penetrated through an enormous number of various section focuses, for example, vehicular databases, remote correspondence advances. New assaults are consistently being created, which implies that automakers will think that it is hard to foresee where programmers will strike. An unbound item made by one of the Original Equipment Manufacturers (OEMs) can furnish programmers with extra passage focuses into a vehicle.
- *Critical risk for drivers and passengers lives*: Regardless of whether only a couple of sensors are misguided or just few ill-conceived messages are sent, a vehicle could encounter glitches that place the lives of drivers, travelers, and people on foot in danger.
- *Improving the security*: When route planning and safety of passengers is considered, information like visibility of road and its conditions, safety index of the road, driving safety status etc... are essential. There needs to be deeper research into the analysis of driving safety with better accuracy. The computing technology these days has made it easy for real time safety analysis. Deep learning technology can be used to predict realtime road safety. These predictions are

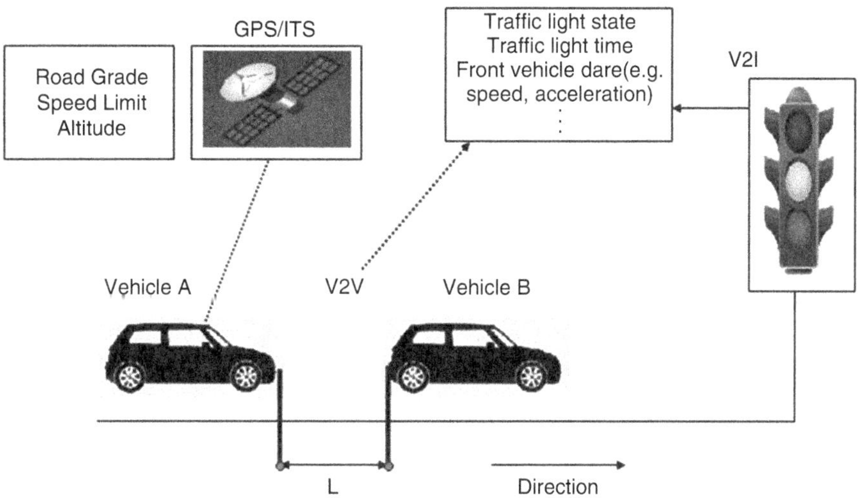

Fig. 17 General V2X infrastructure

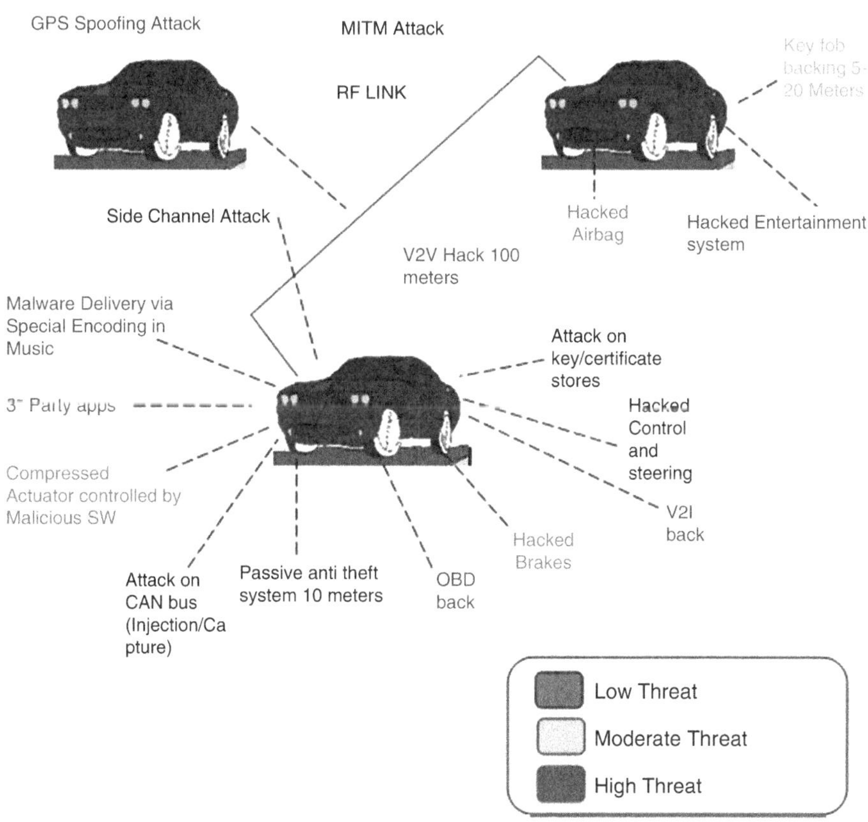

Fig. 18 Possible safety attacks on an AV

based on the data that is obtained from the vehicles trajectories from GPS module and data collected from VANET's through mobile sensing. Deep learning can help equip the driver with the drivers with information regarding the surroundings to prevent road accidents. Another work applied profound figuring out how to improve vehicle security and solace by performing human elements evaluation and showing the encompassing data to the drivers. Giving the encompassing data to the drivers is among one of the different answers for forestall street side mishaps. An ongoing DL-based arrangement is proposed in this work for object location, recognizable proof and acknowledgment of street deterrents in complex rush hour gridlock circumstances. This data handled by one vehicle can be passed to the vehicle behind through V2X to improve the general security factor. Several individuals kicked the bucket in the chain of mishaps brought about by poor perceivability e.g., because of mist, or awful climate. The V2X worldview can assume a vital job in moderating such mishaps also.

Conclusion

This chapter discusses various technical obstacles encountered in the development of autonomous vehicles. The design of the autonomous vehicles has developed to a great extent, from being basic robotic cars to much more efficient and sophisticated/vision guided vehicles. There are still gaps in the research like, there is still lack of accuracy and efficiency in the algorithms, road conditions are still not ideal, dealing with the unpredictable weather is still not possible. Vehicle-to-vehicle communication has been introduced recently and is still in its infancy. Apart from these there are ethical issues, reliability issues, laws, and user acceptance of technology that need to be addressed in the future. At present cars with semi and fully autonomous features have already been launched by majority automobile companies, also almost all cars are expected to be fully autonomous in the near future. In this chapter we outlined some of the key innovations as well as existing systems being employed in the autonomous vehicles, and various technical hurdles in implementing these hardware or software features. Hence, we would like to put these issues forward to discussion.

References

1. D. Elliott, W. Keen, L. Miao, Recent advances in connected and automated vehicles. J. Traffic Transport. Eng. **6**, 109–131 (2019)
2. N. Kalra, Challenges and Approaches to Realizing Autonomous Vehicle Safety (RAND, 2017)
3. M.V. Rajasekhar, A.K. Jaswal, Autonomous vehicles: The future of automobiles, in *2015 IEEE International Transportation Electrification Conference (ITEC)*, Chennai, 2015, pp. 1–6
4. F. Sagstetter, M. Lukasiewycz, S. Steinhorst, M. Wolf, A. Bouard, W. Harris, S. Jha, T. Peyrin, A. Poschmann, S. Chakraborty, Security challenges in automotive hardware/software archi-

tecture design, in *Proceedings-Design, Automation and Test in Europe, DATE* (2013). pp. 458-463. https://doi.org/10.7873/DATE.2013.102
5. Z. Sun, G. Bebis, R. Miller, On-road vehicle detection: a review. IEEE Trans. Pattern Anal. Mach. Intell. **28**(5), 694–711 (2006)
6. A. Reschka, Safety concept for autonomous vehicles, in *Autonomous Driving*, ed. by M. Maurer, J. C. Gerdes, B. Lenz, H. Winner, (Springer, Berlin, 2016), pp. 473–496
7. A. Krizhevsky, I. Sutskever, G.E. Hinton, Imagenet classification with deep convolutional neural networks, in *Advances in Neural Information Processing Systems* (2012), pp. 1097–1105
8. R. Girshick. Fast r-cnn, in *Proceedings of the IEEE International Conference on Computer Vision* (2015), pp. 1440–1448
9. K. Mayilsamy, N. Ramachandran, V. Sunder Raj, An integrated approach for data security in vehicle diagnostics om internet protocol and software update over the air. Comput. Elect. Eng. **71**, 578–593 (2018)
10. A. Tigadi, R. Gujanatti, A. Gonchi, B. Klemsscet, Advanced driver assistance systems. Int. J. Eng. Res. General Sci. **4**(3), 151–158 (2016)

Artificial Intelligence and Sensor Technology in the Automotive Industry: An Overview

S. Meenakshi Ammal, M. Kathiresh, and R. Neelaveni

Introduction

AI is an intelligent tool that is used to simulate human intelligence in machines and mimics their activities like humans in reasoning, learning, perception and problem-solving. In AI, learning is achieved by trial and error, and reasoning is performed by derived inferences which are appropriate to the situation. Problem-solving is a systematic search to reach a predefined goal through some actions. In perception, AI interprets the sensory information which is captured from the real world and extracts the relevant knowledge. For any problem, AI performs many tasks such as aggregating the relevant information, identifying the possible solutions from the information, selecting appropriate solution and creating and reviewing the decision. The specific characteristic of AI is, for any problem, which takes most appropriate action to achieve a solution or a target. AI-based systems use a huge amount of data and intelligent algorithms to mimic the cognitive skills of the human. Machine learning is the subfield of artificial intelligence which builds an analytical model through experience for prediction and decision-making problems. Deep learning is the subfield of machine learning which is built with a multilayer neural network to extract higher-level features from raw data [1]. Figure 1 illustrates the relationship among artificial intelligence, machine learning and deep learning.

Though the AI is being applied across several sectors and industries because of its reasoning ability to learn and solve problems rapidly and independently, the automotive industry is at the forefront of using AI. Recently, manufacturing of self-driving cars which is the most prominent technology in the automotive industry requires advanced software, hardware, communication tools, a high computing and central processor and memory. In AV, the application run in the processor is used to analyse and implement the data received from the sensors and directs the actuators

S. M. Ammal (✉) · M. Kathiresh · R. Neelaveni
PSG College of Technology, Coimbatore, India

M. Kathiresh, R. Neelaveni (eds.), *Automotive Embedded Systems*, EAI/Springer Innovations in Communication and Computing,
https://doi.org/10.1007/978-3-030-59897-6_8

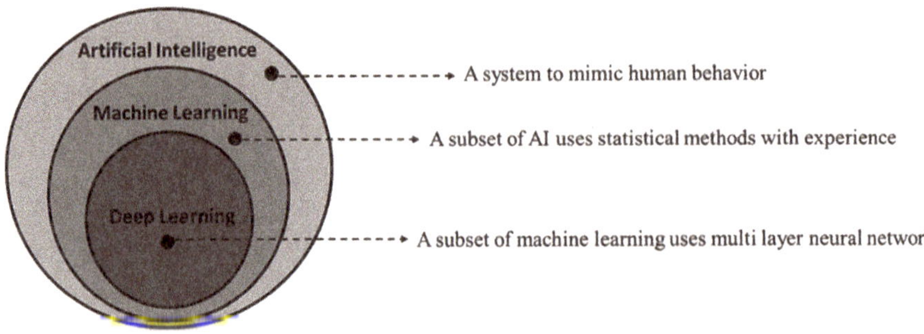

Fig. 1 Artificial intelligence and its subset

to perform specific actions such as applying brakes, changing gears, etc. The sensor components built in the AV have the ability to identify the nearby objects such as road obstacles, pedestrians and cars and help to avoid collisions on the road. Beyond automated features, AI also ensures convenience and safety. Moreover, AV reduces fuel consumption and emission of dust particles and increases new business and market opportunities.

As per the report of the World Health Organization (WHO), each year approximately 1.35 million people die because of road traffic crashes [2]. Most of the road traffic deaths occur because of the vulnerable road users. The report says that road traffic injuries cause the death of people aged 5–29 years and 20–50 million people to suffer from nonfatal injuries and disability. The road traffic injuries also cause economic losses to individuals and their families and consume 3% of gross domestic product of most countries. Low- and middle-income countries encounter 90% of road traffic deaths which is highest in the African region. The US National Highway Traffic Authority [3] has evaluated that the deployment of self-driving vehicles could save USD 300 billion by 2025 because of lessening accidents and safer driving. Manufacturing of AI-based automotive vehicles will cost around USD 10,573.3 million by 2025 [4]. In the automation industry, six levels of vehicle automation [5] are released by the Society of Automotive Engineers (SAE) and the National Highway Traffic Safety Administration (NHTSA) and are often used as a reference point for vehicle automation. Each level is ranked by its advanced automation capability and intelligence as shown in Table 1.

AI-Based Applications in the Automotive Industry

AI has a huge potential to enable various industrial and manufacturing sectors with the latest technologies. In the automotive industry, AI has provided efficient, innovative and safe vehicles to the market. Automobile manufactures have used AI to improve the entire driving experience in diverse ways. AI also has the predictive capabilities which is used for proactive maintenance of the vehicle. This technique

Table 1 Levels of vehicle automation

Levels		Features	Examples
Level: 0	No automation	The vehicles have no autonomous features. The driver performs all actions such as accelerating, steering, etc.	2018 Honda Jazz
Level: 1	Driver assistance	The vehicles have very little autonomous feature. Hence, the driver performs most of the core function	2018 Nissan Sentra
Level: 2	Partial automation	The actions like steering, acceleration, braking and speed maintaining have been automated. The driver performs the activities like driving and environment monitoring	2019 Volvo S60
Level: 3	Conditional automation	The vehicles have most of the automated features except driving. Environment monitoring is also automated and not required by the driver	Honda (under construction)
Level: 4	High automation	The vehicles have all the automated features including driving under some constraints. The driver is the option to control the vehicle	Google's Waymo (under construction)
Level: 5	Full automation	The vehicles have all the automated features with driving functions under all conditions. The driver is the option to control the vehicle	Future project

Source: SAE, NHTSA 2014

uses cloud services to send the sensor data to the manufacturers, insurance companies and advertisers. In the same way, here some of the AI-based applications [6] are given.

AI-Based Driving

Driving assistant: AV built with hundreds of sensors helps to identify the dangerous situations and hazardous conditions on the road. AI helps to enable automated functions like emergency braking, collision avoidance, pedestrian and cyclist detectors, smart cruise directions, blind spot monitoring and cross traffic alerts. Driver-assisted steering helps to take automated actions to avoid accidents.

Self-Driving/Driverless Automobiles Self-driving cars are incorporated with a wide range of technologies such as machine learning, deep learning, gesture control features, computer vision and natural language processing which provide cognitive skills to the vehicles to react according to the situation and to drive safely without a human driver. AV driven by natural language processing responds to voice commands without human involvement and takes required actions to provide safe and convenient driving.

Cloud Services

AI-based vehicles built with sensors generate a huge amount of data. To perform intelligent actions during driving, the connected vehicles (CV) need to send and receive gobs of data via cloud services. AI-based cloud platforms assist to enable the communication in the AV whenever it is needed. In this way, the CV will generate around 4000 GB/day.

Predictive Maintenance The AI-based vehicles have hundreds of sensors which monitor the vehicle's operation continuously and have the ability to predict the failures and disaster before they occur. Hence, the vehicle's owner and the manufacturer can work proactively to prevent component failure.

Individualized Marketing The AI-based cloud platform can also provide location-based information. As AV has the ability to recognize the driver's requirements, it could make suggestions such as the nearest hotels or the nearest gas station.

Automotive Insurance

The insurance companies have collaborated with AI and created a new scheme called Insurtech. This scheme uses the driver's risk factors and make the driver's risk profiles.

AI-Powered Driver Risk Assessment It analyses the drivers' behaviour using their past information and risk profiles and hence could identify the changes in behaviour and activities during driving and could predict the personal issues which causes unsafe driving.

Car Manufacturing

Manufacturing process could also be automated by using AI, and hence it could improve the lifetime of the machinery by 20%. By using AI in the manufacturing process, the machinery inspection cost and the maintenance cost will be reduced, and the quality control will be increased. In automotive plants, AI-enabled automated guided vehicles (AVGs) are used to deliver materials without human intervention. The AVGs have the ability to find out the obstacles in their ongoing paths and help to change the direction accordingly. The AI assists to detect irregularities and defects in materials and to adjust accordingly. The AI-enabled manufacturing process helps to deliver better finished products in a short duration.

Driver Monitoring

AI-enabled cars could help to analyse the driver's behaviour using advanced cameras and IR sensors. Vital signs such as body temperature and blood pressure can also be monitored to track the health condition of the driver. By analysing head position, eye gaze and eye openness of the driver, AI software detects drowsiness of the driver and wakes up the driver without human intervention. During an accident, AI could quickly analyse the driver's posture and upper body position and then deploy the airbags accordingly.

Driver Identification The AI-enabled cars could detect whether the driver is in the proper seating position or not. The AV automatically adjusts the seats and mirrors when the driver gets into the vehicle.

Driver Recognition AV uses facial recognition algorithms to recognize each family member and the driver. Using the facial recognition technique, as the individual family members have their own behaviour and preferences, AI helps to adjust the seat, mirror and the temperature accordingly.

Supply Chain Automation

In recent days, automotive manufacturers have their supply chain throughout the world. They have high impact on the global economy. Even small glitches or breakdowns found in the supply chain will extremely affect the entire process. The AI-enabled supply chain enables to get the entire control over all process which includes planning, manufacturing, inventory tracking, delivery, logistics and management. The AI-driven system could automatically calculate the requests for tools, labour and repairs.

Learning Algorithms in the Automotive Industry

In the autonomous industry, various technologies such as machine learning, deep learning and cognitive computing are integrated with Internet of Things (IoT) to provide innovative and essential features. An electronic control unit (ECU) is a central part which executes various function-specific software to perform the autonomous car's task. Machine learning is the core algorithm running in the ECU. In the AV, the primary function of machine learning algorithms is to monitor and analyse continuously the environment data and to predict even the small changes in the environment. Machine learning algorithms have the ability of self-learning. It needs a huge amount of data to learn and also improve the accuracy from the past experience. Learning includes various methods such as instance-based learning, mapping the input to the output or understanding the data patterns. Different types of machine learning algorithms are applied in the automotive industry based on the requirement of the application [7].

Supervised Learning

The primary goal of supervised learning is to make predictions about future outcomes by analysing the data. Under direct human supervision, the supervised learning algorithms receive data for both input and output and are trained explicitly. The algorithms start its execution to find out the rules for mapping the input to the output. Until reaching the desired level of accuracy, the training process will be continued. The final trained model will predict the desired output. The input data are called training data which always have a known label. Some of the supervised learning algorithms include decision trees, logistic regression, linear regression, random forest, Naive Bayes, support vector machines (SVM) and neural networks. The two types of supervised learning algorithms are classification and regression.

Regression This technique identifies the correlation among the sample data and then makes the prediction of outcomes based on the correlation. It attempts to identify the mapping function from the input variables to continuous output variables.

Classification This technique uses labelled historical data as input and establishes a model using labelled data through the training process. The trained model has the ability to learn the type of object it receives and properly categorize them. It attempts to build the mapping function from the input variables to discrete output variables.

In the self-driving car, the main application of machine learning algorithms is to continuously monitor the surrounding environment and to identify the changes in the surroundings. The regression algorithm is mainly used for predicting the events. In the algorithm, the relationship among the variables is estimated by three metrics such as the type of dependent variables, the number of independent variables and the shape of the regression line. In advanced driver assistance systems (ADAS), the images obtained from the cameras and the RAdio Detection And Ranging (RADAR) are the key inputs for object localization and movement prediction. For example, in the ADAS systems, regression algorithms are used to develop a statistical model which provides the correlation between an image and the location of an object in that image and returns the object location as the output. Some of the regression algorithms used in AV are neural network, Bayesian regression and decision forest. Pattern recognition algorithms are applied to detect the relevant object from the images. SVM and KNN are the most used algorithms for pattern recognition used in self-driving vehicles. Decision matrix algorithms like gradient boosting (GDM) and AdaBoosting are the mainly used algorithms in the automotive industry. These techniques are mostly applied in AV for taking decisions while driving such as taking a left turn or right turn or applying the brake which depends on the situation.

Moreover, they are used in prediction of the next movement of the objects or vehicles.

Unsupervised Learning

Unsupervised learning uses unlabelled data as input to train algorithms and then make predictions. This technique is used to extract the internal structure present in the unlabelled data. It identifies the similarity and hidden patterns in the input data. Mostly, it is used to remove the redundant data or to organize data by its similarity. Examples of unsupervised learning algorithms include k-means clustering, association rule and PCA (principal component analysis). The two types of unsupervised learning algorithms are clustering and dimensionality reduction:

Clustering Clustering is a technique which categorizes the input data based on their internal patterns without any prior information. The clusters are formed based on the similarities of the input data.

Dimensionality Reduction This technique is applied to remove the redundant information and also to retain essential data.

In ADAS, the images from the cameras contain unnecessary environmental data which need to be filtered for recognizing relevant objects. Pattern recognition is an essential part before classifying an object. This technique removes the noise and unwanted data by retaining circular arcs, object edges and line segments of the object. It combines the various features of an object such as the circular arcs and the line segments in various ways to recognize the relevant object. PCA is the most common type of pattern recognition technique.

Usually the images obtained from the sensors are not clear, and so it is difficult to locate and recognize the objects. Hence, the algorithm is not able to classify the low-resolution images properly. The clustering algorithm enables to extract the structure of the data points in low-resolution images. The commonly used clustering algorithms in AV are k-means and multi-class neural network.

Reinforcement Learning

Reinforcement learning (RL) is a self-learning algorithm which employs a continuous cycle of trails and errors in the interactive environment using an agent and learns from its own experiences and its feedback. For solving any problem, the RL takes actions as in a game-like situation. Based on the actions performed, RL generates either rewards or penalties, and the best performance of the RL is achieved by finding a suitable action which should maximize the total reward. The trained model of RL is built with the policies, and the maximum reward is achieved when properly

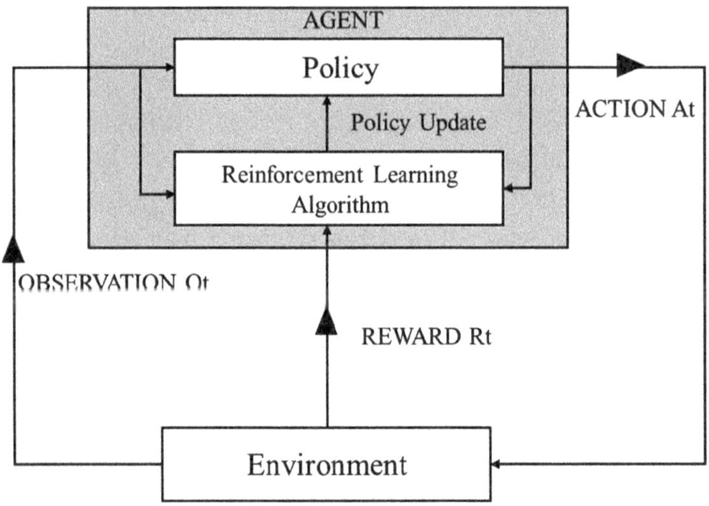

Fig. 2 Workflow of the reinforcement learning

following the policies. Based on the consequence of a current action, the next action will be taken to achieve better outcome. The basic building blocks of RL are environment, state, reward, policy and value. The environment denotes the real world and the state denotes the current situation of the agent. Reward represents the feedback from the environment, and policy is the rule to map the agent's state to actions. Value is the future reward which will be received by taking an action. RL is used in the automotive industry in decision control, braking systems, etc. The workflow of the RL is illustrated in Fig. 2.

Deep Learning

In conventional machine learning methods, the features are extracted from the raw data using feature extraction and feature engineering techniques. Then these features are used by the machine learning techniques to classify the patterns in the input. Feature extraction plays a key role in the conventional methods to extract suitable features. This method is not scalable and also time-consuming for unstructured data. Deep learning (DL) is a subdomain of machine learning algorithm. One of the key properties of DL algorithm is that the model directly extracts the features from the raw input automatically without any feature engineering techniques. Some of the libraries used for training DL model include Keras, TensorFlow, Caffe, Microsoft Cognitive Toolkit, Theano and PyTorch. DL algorithm requires an intensive computation to train a model. DL needs repetitive arithmetic operations, and so it requires multi-core computing techniques. As the normal CPUs have a limited number of arithmetic logic units (ALUs), DL applications use graphical processing units (GPUs). The GPUs are designed with thousands of ALUs and have multiple processors which are connected parallelly. The power of 16 CPUs is equal to the power of 1 GPU. Recently, the DL models are designed with hundreds of millions of weights, multiple hidden layers and billions of connections among units. Some of the companies such as Intel, Mobileye, NVIDIA, Qualcomm, Samsung and Apple have designed and developed specialized chips for real-time deep learning applications. These chips are mostly used in cameras, robots, smartphones and autonomous cars. The most common models of DL are the recurrent neural networks (RNNs) and convolutional neural networks (CNNs). CNNs have the ability to process images, speech, video and audio. RNNs enable to process sequential data such as speech and text [8].

Convolutional Neural Network (CNN)

CNN-based DL models are mostly used to classify and to analyse visual imagery. A typical CNN is built with multiple layers such as an input layer, an output layer and multiple hidden layers. CNN executes a specialized kind of linear operation called convolution for node operations. In case of conventional artificial neural networks (ANNs), a vanilla matrix multiplication is applied as the node operation. The hidden layers of a CNN are built with a series of convolutional layers with filters that convolve with a cross-correlation or other dot product. The CNN architecture is built with three types of layers which are convolutional layer, pooling layer and fully connected layer and they are colloquially called as convolutions. First the convolution layers extract various features of the input image for analysis, and then the fully connected layer uses the output of the convolution layer to predict the image type. The pooling layer is used to retain essential information and bypass the unwanted information generated for each feature. High accuracy is achieved by repeating the process of convolutional and pooling layer.

Recurrent Neural Networks (RNN)

RNN is a type of neural network which is designed to recognize patterns in temporal or time series of data such as handwriting, sensors' data, the spoken word, etc. In traditional neural networks, all the inputs and outputs are independent to each other. As most of the real-time applications have temporal data, RNN is well suited for processing sequence or time series data. In RNN, the output from the previous process is given as the input to the current process. The most important feature of RNN is its capability of hidden layer which is made of memory cells to remember all the processed information of sequence data. Some of the examples of RNN architectures are long short-term memory (LSTM) and gated recurrent units (GRU). LSTM and GRU are built with gates which are used to control the flow of information. These gates have the ability to learn whether the data in the sequence is important or not.

Vehicle Dynamics Virtual Sensing

Automotive stability control systems are implemented based on accurate vehicle dynamic state information. In vehicle motion and vehicle control, the knowledge about certain tire-vehicle dynamic state is considered as a very important input. Vehicle states are estimated by using two factors such as vehicle sideslip angle (β) and the tire-road friction coefficient (μ). Vehicle's stability is estimated using sideslip angle. In the automotive industry, vehicle sideslip angle is estimated using DL techniques. The RNN with a LSTM cell [9] is built to find out vehicle sideslip angle, and the key parameters used for processing are the wheel speeds of the four wheels, the lateral and longitudinal acceleration, the vehicle speed, the front and rear steering angle and the yaw rate. This method has used eight layers of LSTM network and a number of neurons which range from 40 to 100. On the other hand, this method does not include any environmental conditions such as tire-road surface conditions to prove the robustness of this method.

The vehicle sideslip angle is estimated using RNN with the GRU cell [10]. In this method, the outputs of a kinematic model are combined with the inputs of the network. This method calculates the rate of change of vehicle sideslip angle in time. This process has been used 6 million data points which are taken in different road surface conditions such as dry, wet and snowy. The data points are sampled at 100 Hz. The DL methods for estimating vehicle sideslip angle have given better results than the model-based and kinematic-based methods. Still the network is to be optimized for obtaining high accuracy. The other key parameter in the automotive industry is road friction estimation. The LSTM-RNN [11] is designed for measuring dry wetness of the road surface which is estimated from tire-surface interaction. Further, this technique is used to classify tire interaction on wet and dry surface. Moreover, this method hasn't included the tire-road noise and tire wear condition to prove the robustness of the method.

Vehicle Health Monitoring

DL methods are also applied to investigate the working condition of the vehicle parts. DL methods enable to detect the damage of a vehicle automatically. DL methods provide enormous opportunities for the vehicle repair workshops, insurance companies and drivers, and it reduces the cost because of manual labour. A system [12] is designed to detect damages in the outer frame of a vehicle. The OEMs (original equipment manufacturers) facilitate many commercial offerings which integrate vehicle sensors and advanced algorithms to continuously monitor the vehicle battery, fuel pump and starter. The DL methods provide many advanced predictive maintenances techniques to the Fourth Industrial Revolution. The various sensors built inside the AV help to predict the component faults in advance. An algorithm based on CNN [13] is developed to detect defects in the vehicle tires.

Autonomous Driving

Most of the automobile service providers such as Bosch, Toyota, Volvo etc. are working in the construction of AV. The various sensors like RADAR, light detection and ranging (LIDAR), camera, etc. built in AV generate an enormous amount of data for every second. In traditional vehicles, the vehicle control is done using hand-coded decision logic which is a model-based approach. Recently, most of the AVs are designed using end-to-end DL framework which enables to make all vehicle controls based on the situations. In end-to-end DL framework, the neural network space is coded with entire decision-making logic which provides the required driving decisions based on the surrounding environment. NVIDIA corporation provides various applications using CNNs which include path planning, lane making detection and semantic abstraction without any manual decomposition. Apart from vehicle control, end-to-end DL model enables driver monitoring and driver assisting tasks.

Functional Architecture of an Autonomous Vehicle

The functional architecture of an AV represents various software and hardware components and its interaction to each other for executing the task [14]. In AV, the hardware part includes sensors, actuators and connected vehicles. The software part includes perception, planning and controlling. Figures 3 illustrates the functional architecture of an AV. Figure 4 shows the sensors in the autonomous vehicle.

Sensors The sensors obtain the input from the environment as a raw data. The most commonly used sensors in the AV are LIDAR, camera, GPS and RADAR. The data from various sensors are combined using a technique called sensor fusion.

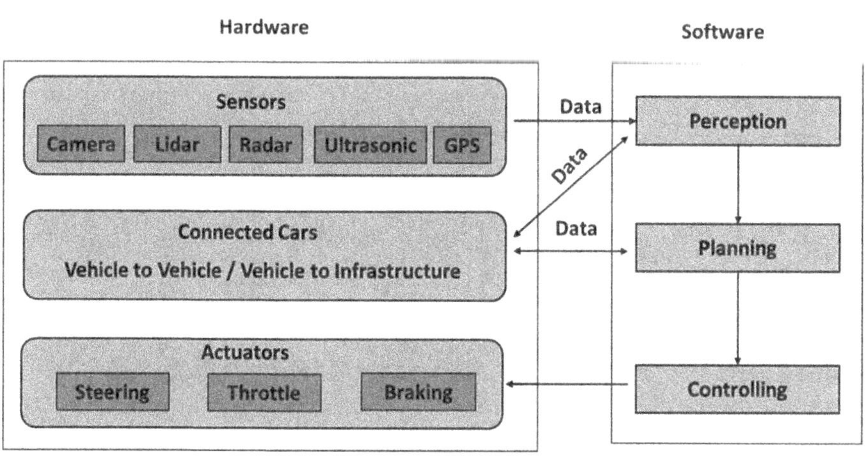

Fig. 3 The functional architecture of an AV

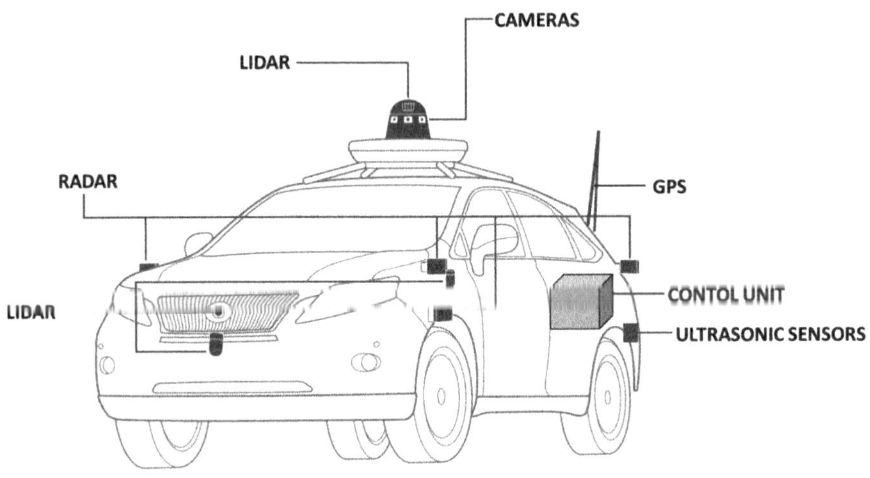

Fig. 4 Sensors in the autonomous vehicle

Connected Cars This represents the communication between a vehicle and an object. The object may be a car, a component or an environment. The two types of connection in AV is vehicle to vehicle and vehicle to infrastructure.

Actuators Actuators are the mechanical components present in the AV. They are responsible for the physical operation of the AV.

Perception The perception represents the ability of the AV to extract the useful knowledge from the raw data of sensors and from the connected vehicles. It interprets the information obtained from the environment and generates useful data such as obstacle and lane marking detection, etc. Localization indicates the position of an object in the environment.

Planning The planning represents the actions that are to be taken for accomplishing some tasks based on the information obtained from the perception. It takes certain decisions such as how to navigate in the environment.

Controlling The controlling refers to the mechanical operation performed based on the input from the planning system. It represents the operation of vehicle such as moving, applying brakes, etc.

Role of Sensors and Actuators in Autonomous Vehicles

AV is constructed by a combination of various technologies such as sophisticated algorithms, communication tools, sensors, actuators and high computing processors. In the automotive industry, sensor technology enables to develop a wide range of applications in safety, entertainment, traffic control, navigation and guidance. Hundreds of sensors are fixed inside the vehicle to capture information about the environment surroundings. Recently, tire pressure sensor and rear-view visibility systems are the mandatory components in the vehicles. The sensors are used to reduce traffic congestion and to provide road safety. Most of the sensors installed in the vehicle are used to provide assistance, convenience, passenger's satisfaction, etc. The average number of sensors in the AV is around 60–100. When adding new features and applications in the AV, the number of sensors installed in the vehicle will also increase, and it will be around 200 sensors inside a vehicle. Based on the applications, the sensors fixed in the vehicle are categorized into five as follows: sensors for diagnostics, sensors for safety, sensors for environment monitoring, sensors for convenience, sensors for driver monitoring and sensors for traffic monitoring [15]. Table 2 illustrates the various types of in-vehicle sensors and its applications. Figure 3 depicts the mostly used sensors in AV.

The sensors and actuators used in the AV are broadly classified into three categories such as

1. navigation and guidance,
2. safety and driving and
3. performance.

Navigation and Guidance

Navigation and guidance systems are constructed with inertial measurement unit (IMU) which contains sensors such as accelerometer, gyroscope and magnetometer. The IMU estimates the vehicle velocity and orientation. The system is combined with global positioning systems (GPS) to obtain current position of the vehicle. Though GPS is very essential for AV, the signals from the GPS may get disturbed by tunnels, tall buildings, etc. Hence the inertial sensors provide rotational and linear motion of the vehicle. Microelectromechanical system (MEMS)-based IMU is used in the AV. The system enables various services such as to detect the users' location, to find the optimum route to the destination and to avoid unwanted delays while driving. GPS also uses cameras, RADARs and lasers to estimate the environment conditions. The RADARs are used to detect the road conditions such as snowfall, heavy rains, etc. The lasers fixed in the roof of the vehicle provide three-dimensional view of the surroundings. The application provides various information such as road conditions, driving routes and traffic conditions which could assist the drivers to take fast decision while driving. GPS, RFID and magnetometers could help to collect information about parking places. The applications which are developed to help drivers in unknown and tourist places use the sensors such as RADAR sensors, inductive loops, weather sensors and cameras which are used to find out the important places in a city and the route to reach destination. Moreover, the applications help to know

Table 2 Sensors in the autonomous vehicle

Type of application	Sensors	Description
Diagnostic	• Air bag sensors • Gas composition sensors • Pressure sensors • Position sensors • Temperature sensors • Chemical sensors • Inertial sensors	Estimate the performance of the vehicle while driving and detect the vehicle failures
Environment	• Cameras • Pressure sensors • Ultrasonic sensors • Humidity sensors • Temperature sensors • Rain sensors	Give alerts based on environmental conditions and automatically adjust the temperature
Traffic	• Proximity • Ultrasonic • RADARs • Cameras	Detect the traffic condition using spatial and temporal information
Safety	• Proximity sensors • Cameras • RADARs and laser beams • Micromechanical oscillators • Wheel speed sensors • Haptic sensors • Night vision sensors • Ultrasonic sensors • Inertial sensors	Monitor surrounding environments and prevent disaster and accidents in advance
Assistance	• Humidity sensors • Temperature sensors • Gas composition sensors • Torque sensors • Fogging prevention sensors • Ultrasonic sensors • Position sensor • Rain sensors • Image sensors	Provide comfortable, secure and convenient driving
Driver/user	• Thermistors • Cameras • Heart rate sensors • Electrocardiogram (ECG) sensors • Electroencephalogram (EEG) sensors	Monitor the health condition and behaviour of the driver and measure the performance of the driving

the real-time traffic conditions and weather conditions and suggest alternate routes and give the travel time to reach the destination.

Safety and Driving

Safety applications focus on reducing the number of accidents, fatalities and injuries. The lane management application uses cameras which are mounted behind the rear-view mirror and monitors the road lane markings and finds out any drifts outside the lane. RADAR and speed and distance sensors are being used in adaptive cruise control applications to control the vehicle speed and to maintain a constant distance from the object in front. AV has a 360° view of the surrounding environments to monitor the obstacles and other vehicles on the road. The safety application is provided through various assistive functions such as adaptive cruise control, lane assistance, emergency assistance and light assistance. Though an array of cameras helps to capture images in real time, it requires heavy graphic processing and high-resolution pictures. To overcome the problem, LIDAR is fixed in the roof of the vehicle which provides 3D graphics of the surrounding environment. It helps in various ways such as to reduce vehicle collision and to detect obstacles and other objects. The RADARs fixed in the front and rear bumpers of the AV are very useful in many applications such as parking control, lane changing and bumper-to-bumper traffic. The IMU is also used to predict the risk factors which cause vehicle disasters. RADAR and laser sensors are used to monitor the road and to prevent collisions. If something is found nearest to the vehicle, the sensors used in the vehicle direct the control to apply brakes and to adjust throttle automatically. The application of blind spot detection uses RADAR to detect blind spot zone and gives alerts to the driver accordingly. The collision warning system uses position and speed sensors to predict the vehicle collision and sends a warning signal in advance. Data fusion techniques and intelligent algorithms are also being applied to further improve the applications to obtain fast response and to make fast decisions during disaster time. The sensors used in AV monitor the driver's performance and behaviour such as alcohol consumption, fatigue and emotional state. The warning systems monitor the eyes and head motion of the drivers using cameras and detects drowsiness. Steering angle sensors are used to detect anomaly conditions during driving and give alerts as audio or vibration signals. Front-facing video cameras are used to monitor the left and right lanes and give an alert to the drivers if the vehicle drifts from the lanes, thus reducing the chance of an accident. The cameras are also used to identify the fatigue of drivers, and alerts are given as a sound signal and a flashing message using control panel of the AV. Driver's health condition and vital signs such as body temperature, respiration rate, breathing rate and blood pressure could be measured using sensors such as silicon photodiode, thermopile and optical and infrared sensors. These sensors deployed on seats and steering wheel help to detect the driver's health and activate the associated process automatically. The driver's depression, stress and emotions could be detected using sensors such as electromyogram (EMG), respiration sensors and electrodermal activity (EDA), and the data obtained from these sensors could be analysed using machine learning algorithms, neural networks and cloud computing.

Performance Management

Designing an AV with hundreds of sensors, actuators and dozens of circuit boards has created issues in power and fuel management. Hence a number of subsystems, unique circuit boards, are fixed inside the AV to utilize power and fuel consumption efficiently and to control thermal dissipation. Voltage consumption of a battery is to be measured to maintain proper vehicle functions. Voltage drop in the batteries is effectively measured using milliohm meters along with differential amplifiers. The components of an electronic fuel injection system such as actuators, step motors and solenoid valves enable to utilize the fuel efficiently [16]. Using different types of sensors, the failures found in the mechanical components could be detected automatically. The functionality of an engine and different parts of the vehicle could be checked using powertrain sensors. Chemical sensors are used to check the quality of the fluid. Temperature sensors are used to check the temperature of fluids and gas. The defects found in the chassis systems are detected using chassis sensors. The engine combustion is found out using composition sensors. In the AV, each vehicle part could be monitored continuously, and the breakdowns could be prevented in advance.

Actuators in AV

The actuators are broadly classified into three based on the operations such as throttle actuation, brake actuation and steering actuation. Recently, most of the AV have been designed with all the actuators which help to control the vehicle in any situation. Electronic control is used to achieve throttle actuation. Electronic stability control is used to obtain brake actuation. The NHTSA has recommended to design vehicles with brake actuation which is given as a mandatory requirement to the automotive industry. Steering actuation is achieved using electric-assisted power steering which removes the use of hydraulically driven power steering system. All the revolutionary changes in the automotive industry have been achieved by the sensors, actuators, algorithms and processors that provide the ability to the AV to understand the surrounding environment and to take decisions automatically without human intervention.

Limitation of In-Vehicle Sensors

Sensors, actuators, processors and control unit are equipped inside the self-driving cars to support many applications and features. Still, there are many challenges in designing an AV. Currently, ADAS and partial autonomous vehicles use camera systems which are based on CMOS image sensors and RADAR. However, the environmental conditions and various lighting conditions make difficulties in detecting objects and capturing images. Hence, low-resolution images reduce the performance of the image processing and acquisition approaches in real time. In LIDAR-based applications, designing a high-resolution LIDAR system in affordable cost is still at the pre-development stage. The LIDAR system uses light spectrum waves to detect objects in the lane and provides three-dimensional images. Still bad weather conditions affect the clarity of the images. The size of the sensors and its integration inside the AV has become a great challenge in the manufacturing process. The RADAR is used to measure the distance between a vehicle and an object and also used to maintain a constant distance from the objects. However, the RADAR could only detect the object in the centre of the road because of its limited ability of vision. As there is no unique assistance system in designing an AV, multi-sensors are integrated to provide various features of the vehicle. For example, camera systems are combined with RADAR systems to detect the obstacles and moving objects in the road.

Connected Vehicle

A connected vehicle (CV) represents the communication of vehicles with each other and with the environment surroundings using specific applications, technologies and services. A navigation guidance system is designed with CV features to enable dynamic route guidance. The AV integrated with the GPS obtains the real-time traffic congestion through cellular signals. A CV embedded with many in vehicle communications tools and networks enables connection with internal devices, external devices, applications, services and networks [17, 18]. A CV facilitates many applications such as parking assistance, infotainment, safety, telematics and remote diagnostics. Interaction with other vehicles is enabled using advanced driver assistance systems (ADASs) and cooperative intelligent transport systems (C-ITS). Vehicle-to-vehicle (V2V) and vehicle-to-infrastructure (V2I) communication has the ability to obtain situation awareness and is hence used to reduce traffic accidents. The V2V and V2I technology allows transportation agencies to access vehicle information without compromising privacy data. This enables transportation agencies to manage the traffic flow and to assist in vehicle disaster on the road [19].

The communication technology used for V2V and V2I is based on IEEE wireless access in vehicular environment (WAVE) standard. This includes dedicated short-range communications (DSRC), the IEEE 1609 family for upper layers and the

IEEE 802.11p for PHY and MAC layers. By using DSRC, the components in the AV could exchange the information within the range of 300 m. Dynamic Spectrum Access (DSA) is another technology provided by DSRC to exchange information in a CV. The Society of Automotive Engineers (SAE) has created a communication standard called J2735 for AV.

Mostly, V2I and V2V communication are performed by 4G/LTE technology which enables communication among passive magnetic sensors, inductive loops and pneumatic tube sensors. The 4G/LTE technology have many features such as high data rate, large coverage and low latency. The applications such as road safety and traffic analysis utilize WiMAX technology to provide coverage up to 50 km in the data rate of 70 Mbps. A global low-power wide-area (LPWA) network is used for transmitting data from different sensors to various components inside a vehicle. Advanced wireless technology which has latency time less than 5 ms is mostly applied in safety applications. High infrastructure and service cost are the main disadvantages of using wireless network technology in the automotive industry. Table 3 shows the various communication protocols and networks used in V2V and V2I applications.

Challenges in AV

The main challenges of sensing systems are blurry transit lines, fast-moving object detection, risky road conditions such as holes in the road and environment conditions such as road flood. The different type of data from independent data sources should be integrated to produce appropriate response in real time. The sensor fusion techniques are required to aggregate a variety of sensor data. Hence, they will help to obtain fast response time and to avoid potential collision situations. In AV, the intelligent algorithms should be designed and developed to process data from various sources through different communication media such as wired/wireless links. The algorithm should have the ability to remove abnormal data and redundant features.

Theft of Data As the CV generates a huge amount of data from in-vehicle sensors, there are more chances for data hackers to steal personal information, financial information, location data and entertainment preferences.

Table 3 Applications and its related communication protocols and networks

Applications	Communication protocols and networks
Driver monitoring	Local Interconnect Network (LIN) Time-triggered lightweight protocol (TTP/A)
Environment monitoring	Controller area network bus (CAN-B), J1850
Performance monitoring	Controller area network bus (CAN-B)
Multimedia and infotainment applications	Media oriented system transport (MOST) Bluetooth, FlexRay, Zigbee, Wi-Fi, Ultra-wideband (UWB)

Theft of Vehicle Recently the digital keys, mobile applications and wireless key fobs are used to lock and unlock the vehicles. These virtual keys could be accessed by unauthorized person through intercepting the communication between a wireless key fob or a vehicle.

Network Risks In V2V and V2I, the communications through Wi-Fi, cellular networks, Bluetooth, etc. have created an easy target for hackers and have more chances of getting corruption in service and network access.

High Cost In object detection method, 360° vision is achieved by integrating various devices. However, it generates a large amount of data and also requires high-power processors to analyse the images. These overall requirements have increased the system cost. Low-cost hardware and communication techniques are the major requirements for the commercialization of AV.

Object Detection The main challenge of AV is to detect fast-moving objects on the lane.

DL Architecture Designing a specialized computing DL architecture for AV is a crucial task in the automotive industry.

Threats There is more chance for data manipulation by unauthorized people, data corruption, device hacking and counterfeiting.

Conclusion

Machine learning and deep learning algorithms play a key role in the functionality of an AV. However, DL models for AV require a huge amount of data and need new infrastructures and tools for computation. Mostly, computer vision problems and vehicle motion control have been solved using existing DL model. In the future, specific computational architecture, unique sensor components and communication technology will be able to design the highest level of semi-automatic and fully automatic vehicles. The usage of sensors in AV enables to create various applications such as driver monitoring, traffic control, environment monitoring, vehicle maintenance, etc. However, the sensors used in AV have its own limitations which should be resolved in the future. Further, AV should find solutions to the problems such as high fuel prices, high emissions of CO_2, traffic congestions and low-quality roads. In the future, AV will arise many ethical issues such as unemployment of drivers. It replaces the manual repairing and so reduces the source of earnings. Modern AV should ensure to guarantee data confidentiality and integrity. It should facilitate to protect data transport and storage and ensure platform integrity. Hence the integration of various technologies will broaden the research area in the automotive industry.

References

1. H. Khayyam, B. Javadi, M. Jalili, R.N. Jazar, Artificial intelligence and internet of things for autonomous vehicles. Nonlinear Approaches Eng. Appl. **2020**, 39–68 (2020). https://doi.org/10.1007/978-3-030-18963-1_2
2. World Health Organization, Road traffic injuries (2020)
3. P. Wadhwani, S. Kasnale, Artificial intelligence (AI) in automotive market, 2020-2026 Forecasts, Global Market Insights. Report ID: GMI3199 (2019)
4. Automotive Artificial Intelligence Market, Report Code: SE 5533 (2017)
5. National Highway Traffic Safety Administration (NHTSA), Automated Vehicles for Safety (2014)
6. S. Zoria, The place of machine learning and artificial intelligence in the automotive industry, Data Driven Investor (2019)
7. J. Li, H. Cheng, H. Guo, S. Qiu, Survey on artificial intelligence for vehicles. Automot. Innov. **2018**, 2–14 (2018). https://doi.org/10.1007/s42154-018-0009-9
8. K.B. Singh, M.A. Arat, Deep learning in the automotive industry: recent advances and application examples (2019)
9. J. Ghosh, A. Tonoli, N. Amati, A deep learning based virtual sensor for vehicle sideslip angle estimation: experimental results. SAE Tech. Pap. **2018**, 7148–7191 (2018). https://doi.org/10.4271/2018-01-1089
10. T. Graber, S. Lupberger, M. Unterreiner, D. Schramm, A hybrid approach to side-slip angle estimation with recurrent neural networks and kinematic vehicle models. IEEE Trans. Intell. Vehicl. **4**(1), 39–47 (2019). https://doi.org/10.1109/TIV.2018.2886687
11. I. Abdic, L. Fridman, E. Marchi, D.E. Brown, W. Angell, B. Reimer, B. Schuller, Detecting road surface wetness from audio: a deep learning approach, in *23rd Int. Conf. Pattern Recognition (ICPR)* (2016), pp. 3458–3463. https://doi.org/10.1109/ICPR.2016.7900169
12. UVeye Vehicle Inspection Systems - Smart Mobility, UVeye - Vehicle inspection systems (2019). https://www.uveye.com/smart-mobility/
13. Y. Zhang, X. Cui, Y. Liu, B. Yu, Tire defects classification using convolution architecture for fast feature embedding. Int. J. Comput. Intell. Syst. **11**(1), 1056–1066 (2018). https://doi.org/10.2991/ijcis.11.1.80
14. O.Ş. Taş, F. Kuhnt, J.M. Zollner, C. Stiller, *Functional system architectures towards fully automated driving*, in IEEE Intelligent Vehicles Symposium (IV) (2016). https://doi.org/10.1109/IVS.2016.7535402
15. J. Guerrero-Ibanez, S. Zeadally, J. Contreras-Castillo, Sens. Technol. Intell. Transport. Syst. **18**(4), 1212 (2018). https://doi.org/10.3390/s18041212
16. S. Abinesh, M. Kathiresh, R. Neelavenik, *Analysis of multi-core architecture for automotive applications*, in International Conference on Embedded Systems (ICES) (2014). https://doi.org/10.1109/EmbeddedSys.2014.6953094
17. C. Varun, M. Kathiresh, *Automotive ethernet in on-board diagnosis (Over IP) & in-vehicle networking*, in International Conference on Embedded Systems (ICES). https://doi.org/10.1109/EmbeddedSys.2014.6953168
18. K. Mayilsamy, N. Ramachandran, V.S. Raj, An integrated approach for data security in vehicle diagnostics over internet protocol and software update over the air, in *Computers and Electrical Engineering* (2018), pp. 578–593. https://doi.org/10.1016/j.compeleceng.2018.08.002
19. D. Elliott, W. Keen, L. Miao, Recent advances in connected and automated vehicles. J. Traffic Transport. Eng. **6**(2), 109–131 (2019) https://doi.org/10.1016/j.jtte.2018.09.005

Advanced Driver Assistance Systems (ADAS)

Maria Merin Antony and Ruban Whenish

Introduction

A driver is one of the "best sensors in the vehicle" and is the main responsible for avoiding crashes. But still, a large proportion of crashes are attributed to the driver errors. A survey was conducted to identify the critical reason for each crash, and the national sample of US crashes from 2005 to 2007 was examined [1]. It was noted that the driver error was the critical reason contributing to 94 percentage of crashes as shown in Fig. 1. These errors included recognition errors, decision errors, performance errors, and nonperformance errors. Recognition errors are the result of inattention and inadequate surveillance of the driver; decision errors arise due to the misjudgments of the driver; performance errors arise due to overcompensation, poor directional control, etc.; and nonperformance errors arise due to sleeping and fatigue [2].

The development and deployment of the new in-vehicle technologies to counteract these driver errors and hence to support the driver to prevent crashes is ongoing. Advanced driver assistance systems (ADAS) are a group of vehicle technologies that warn the drivers timely regarding the risky or hazardous situations to avoid crashes. Some ADAS technologies actively and automatically intervene to avoid hazardous situations or when the system detects that a crash is imminent.

ADAS technologies are the precursor to autonomous vehicles and, depending on the combination of ADAS equipment installed in a vehicle, can allow level 1 to level 2 autonomous driving at the present time as represented in Fig. 2 (SAE International 2014).

M. M. Antony (✉)
Centre for Optical and Laser Engineering, School of Mechanical and Aerospace Engineering, Nanyang Technological University, Singapore, Singapore

R. Whenish
Sri Krishna College of Technology, Coimbatore, India

M. Kathiresh, R. Neelaveni (eds.), *Automotive Embedded Systems*, EAI/Springer Innovations in Communication and Computing,
https://doi.org/10.1007/978-3-030-59897-6_9

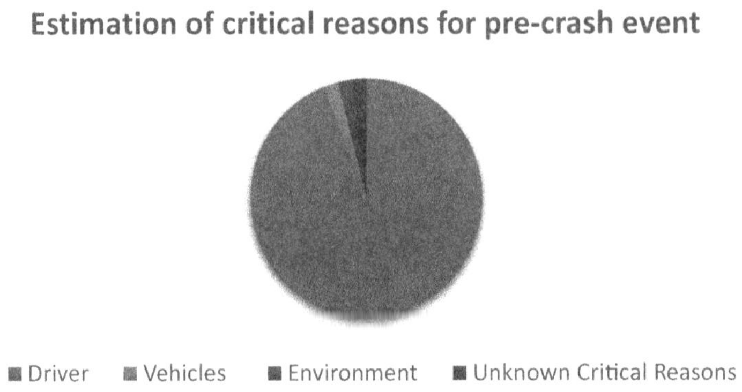

Fig. 1 Estimation of the critical reasons for pre-crash event. Data source: NMVCCS 2005–2007

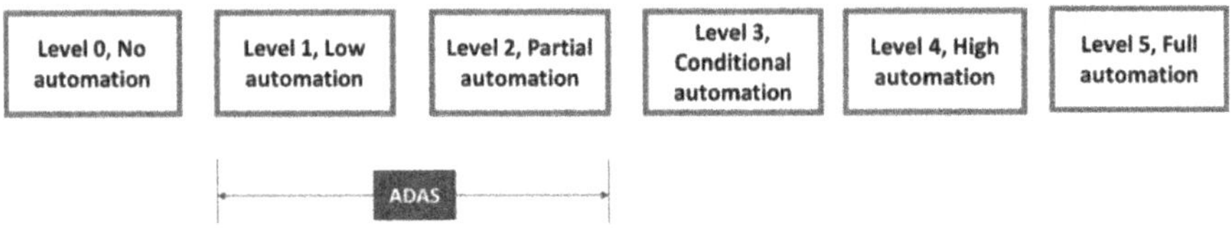

Fig. 2 Levels of autonomous driving as per the SAE classifications and ADAS role in the automation

ADAS Technologies

Driver assistance systems (DAS) were developed aiming at the development of systems that are designed to assist and improve the comfort level of the driver while driving. Initial systems developed as driver assistance systems include parking assistance system (PAS) and cruise control (CC). With the advancements in DAS technologies, the focus got shifted on to the increased level of safety of both the driver and pedestrians. Such advanced driver assistance systems (ADAS) refer to electronics, light-based, or sound-based systems that are integrated together to the development of technologies that automate, facilitate, and improve systems in the vehicles in order to assist drivers for better and safer driving. There are numerous components now developed to be known as ADAS technologies. A complete overview of the features of ADAS technology is shown in Fig. 3.

Various ADAS technologies in a vehicle are realized with the help of many types of environment sensors such as radars, lasers, ultrasonic sensors, and cameras. These systems are equipped with control algorithms that signal the drivers or directly control the brakes or throttles. The control systems of the vehicle with ADAS will behave based on the signals from the environment and according to the state of the vehicle such as steering angle, GPS data, etc. or state of the driver. The ADAS technologies used in modern cars are described in short as below:

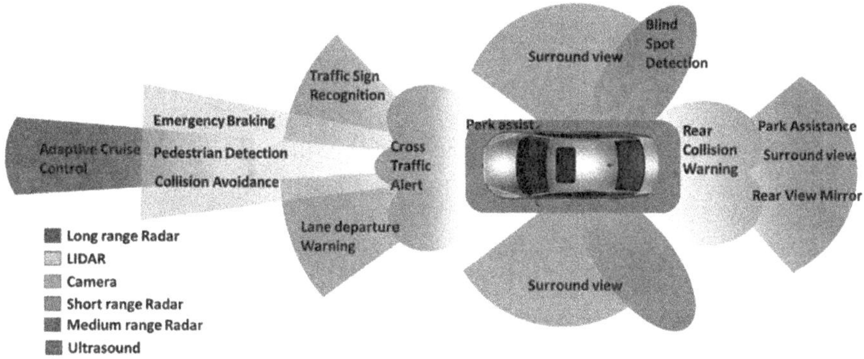

Fig. 3 Overview of ADAS technologies used in a vehicle

Adaptive cruise control (ACC) also known as radar cruise control is a system developed to control or adjust the speed of vehicle to maintain an optimum distance between the vehicles with the help of a radar or a laser system. In ACC, three radar sensors are needed, two short-range radars used to detect objects in the adjacent lane and one with a long range used to detect objects in-path.

Forward collision control with autonomous emergency braking alerts the driver when the speed of the vehicle can result in a collision with a vehicle or an object in front. Autonomous braking (AEB) will prevent collision by applying brakes when there is a high risk of collision and the response of the driver is inadequate to avoid the collision.

Lane departure warning systems give warnings to the driver while there is a risk or straying across the lane markings. The feature uses the long-range radar sensors for the sensing purpose. A Lane estimation algorithm is used to gather the information for lane departure warning (LDW) and lane keeping rely features for ADAS technologies.

Side- and rear-view assistant systems monitor surrounding areas to warn the driver for the blind zones with the help of cameras or radar sensors.

Traffic sign recognition displays are seen on the instrument panel reminding drivers of the current speed limit. The traffic signals are recognized with the help of the same camera which recognize the speed limits. Even the navigation system of the vehicle can be supported by storing the information regarding the speed limits in unsigned roads.

Blind spot detection (BSD) helps the driver to view the blind spots while overtaking. A short-range radar is used for monitoring the road area behind and warning the driver while overtaking the vehicles.

Rear cross traffic alert (RCTA) uses two radars that can monitor at 120 degrees and help in detecting accidents while reversing out in the parking space.

Pedestrian detection technology monitors and detects the pedestrians in the road and provides the driver the necessary safety guidelines with the help of normal and IR cameras.

Fig. 4 Steps taken in the ADAS technology

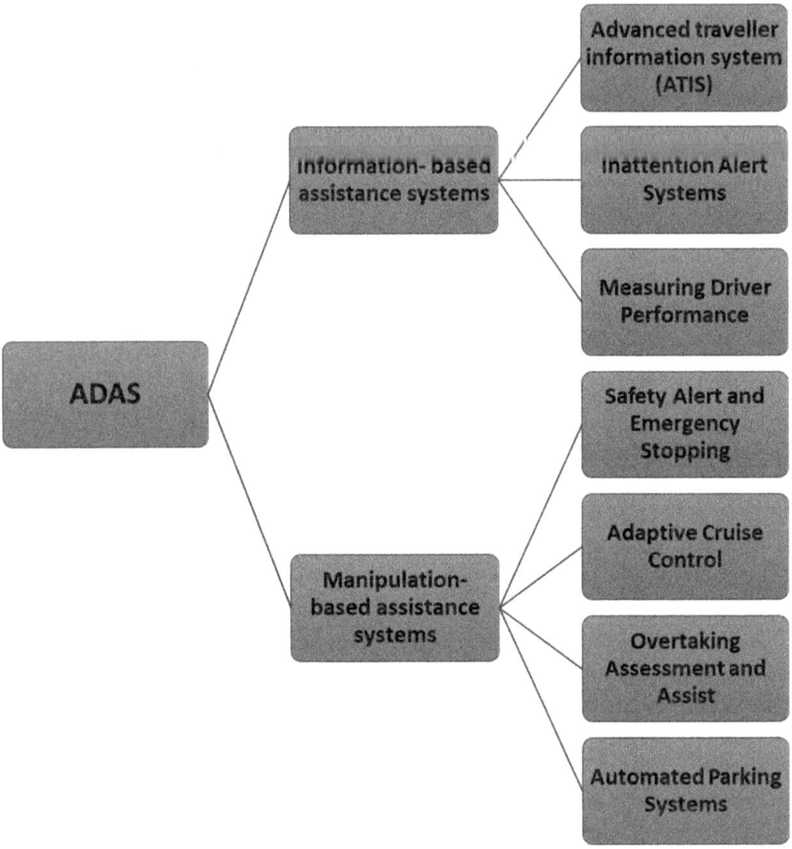

Fig. 5 Classification of ADAS technologies

In each ADAS technology, there are three steps taken as shown in Fig. 4.

1. Recognition
2. Judgment
3. Operation

The first step in the ADAS technologies is recognition with the help of sensors such as radar, LiDAR, cameras, and ultrasonic sensors. Based on the signals from the sensors, the control action is decided with the help of advanced imaging processing or machine learning algorithm.

ADAS systems are basically classified into information-based assistance systems and manipulation-based assistance systems [3]. The detailed classification is shown in Fig. 5.

Information-based assistance systems provide only necessary information and warnings to the human driver. They don't implement any driving decisions and physically operate the vehicles by small amount.

Manipulation-based assistance systems only operate on the vehicle in case of emergency or when asked by the driver.

Information-Based Assistance Systems

During driving, a lot of concerns are present in the mind of a driver regarding the correct route to be taken, the level of traffic congestion, and time delay in reaching the destination on the right time. The information-based assistance systems address all these questions, and such a system do not control the vehicle by itself. The information-based assistance systems are further classified as Advanced Traveller Information Systems (ATIS) as shown in Fig. 5.

The ATIS system performs main functions such as dynamic rerouting and anticipation of traffic congestion and provides information about the surroundings. These information helps the driver to take key decisions regarding the route to be taken. The data for dynamic and historic information about the traffic is made available with the help of the cameras at different locations at various roads. Such information is used to anticipate the congestion and reroute the traffic.

Drivers are distracted by many expected and unexpected reasons such as sleep, tiredness, fatigue, etc. These reasons can be identified by various indicators such as eye movements (blinking frequency of the eyes), head movements (yawning), biological and neurological signals (such as EEG and ECG), and facial expressions (movement of the eyes, lips, etc.)

The indicators of inattention such as distraction are measurable. The effects of distraction can be measured based on movements of the eye, head, and face and signals obtained from the driver's body such as electroencephalography (EEG) and electrocardiogram (ECG). EEG signals carry information about the brain waves and brain activities. In contrast, the ECG signals gives information on heart changes which can be used to interpret and measure the change in the thought process of the human driver.

The general procedure for detecting inattention uses the machine learning techniques. A procedure which is typically followed for the pattern recognition system is shown in Fig. 6.

As the initial step, the noise is removed from the data collected. Data is then segmented, and the feature is extracted for training and to create the machine learning model. The real-world data is then segmented, and the feature is extracted and classified based on the training data and model for the driver to take decision.

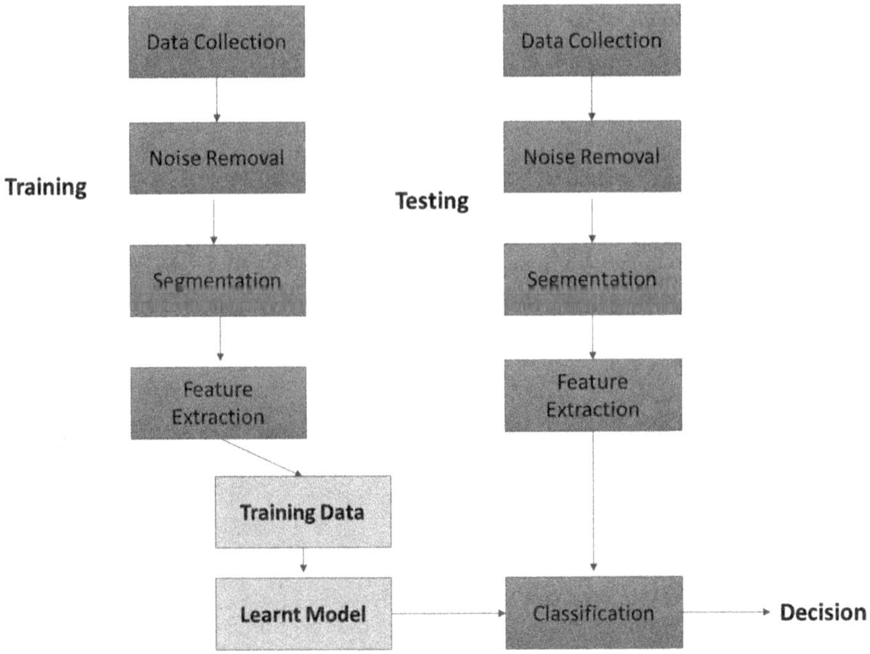

Fig. 6 Procedure used for pattern recognition problem

While detecting the images, the images are reduced in dimensionality using features like scale invariant feature transform (SIFT) and speeded up robust transform (SURF). Another possible approach is to use color filters for the detection of the face. There are numerous techniques that are used for face detection and recognition. Modelling of blinking of the eyes, gaze, and head pose was recognized in order to identify the inattention and fatigue in the drivers during driving.

The next task is to extract the features from the complete image, or signal is broken down using dimensionality reduction techniques and then can be used for classification. The techniques such as principal component analysis (PCA), independent component analysis (ICA), and linear discriminant analysis (LDA) are used for dimension reduction and extraction. Later, the classification is done with the help of popular classifiers such as neural networks, support vector machines, fuzzy inference systems, adaptive boosting algorithms, and decision trees.

Hybrid techniques are also used which include the combination of any of the above discussed techniques (Figs. 7, 8, and 9). The commonly used multiple modalities are discussed as follows:

(a) Data level fusion: Here the fusion is performed on the raw data of the different modalities. The fused data is then passed through feature extraction and classification in order to help the driver make an appropriate decision.
(b) Feature level fusion: It is fused at the feature level, and then classification is done.

(c) Score and decision level fusion: A score function is generated after classification from each data source by an integrator based on which decision is made. Here, the generated score function is fused to form the final decision.

Measurement of the driver performance is important for transportation authority to access the driver performance to extend or cancel license to drive. The systems for measuring driver performance are built inside the vehicle to continuously rate the driver's performance. The best way to indicate the driver's performance is by assessing the distances of the vehicle in front, fluctuations in speed, and the ability to make smooth lane changes and to maintain the lanes. Other parameters that can be used to measure the condition of driver are the way of handling the steering wheel, reversal speeds of steering, and pressure distribution on the seat of the driver. All these parameters are helpful in detecting fatigue and drowsiness of the driver which are continuously monitored by the ADAS technologies in the car.

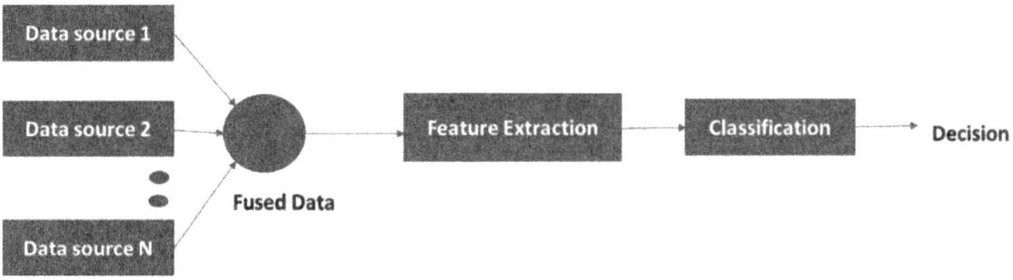

Fig. 7 Data level fusion

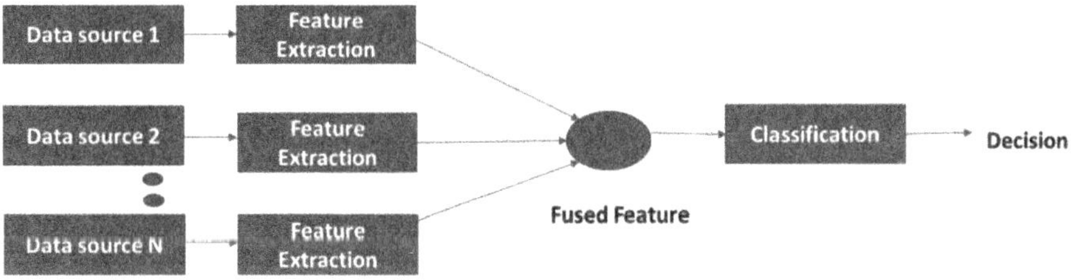

Fig. 8 Feature level fusion

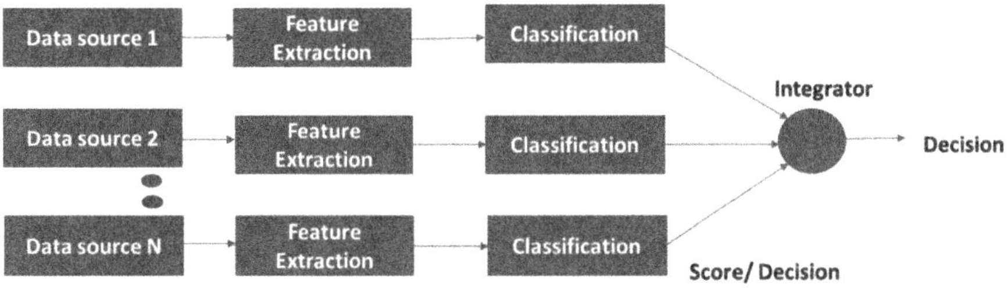

Fig. 9 Score and decision level fusion

Manipulation-Based Assistance Systems

Manipulation-based assistance systems not only detect the state or risk involved in driving but also manipulate the vehicle on behalf of the driver. These systems come into action in three situations as follows:

- When emergency actions are needed and there is no time to warn the driver
- To perform certain regular and simple tasks which do not require human wisdom
- To perform certain precise and specialized tasks that are hard for humans to perform such as overtaking and parking

There are situations where the alerting or warning of the driver is ineffective due to time constraints and there is a need of quick response in some situations to take safe action. Such situation needs to be addressed by emergency braking systems. In contrast, safety alert systems predict the risky situations and provide warnings to take necessary action. The collision avoidance system is one of the common examples to explain these systems.

Another similar problem is the collision with pedestrians, especially the old-age people and children who do not take much care while crossing the roads. There are pedestrian detection systems which detect and alert the driver to avoid collision and risks.

The adaptive cruise control (ACC) maintains the speed of the vehicle constantly by the use of brake and throttle. The driver is relieved/freed from continuously monitoring the speed by the use of ACC, and he just needs to control the steering. The sensors such as radar, LiDAR, or ultrasonic sensors will continuously monitor the distance from the vehicle and environment. While vehicle following is successful with no risk and accidents, the ACC follows an elastic band. This band is always maintained to be smooth and collision-free and tends to be the path for trajectory for the vehicle to be followed as shown in Fig. 10.

Overtaking is desirable and helps the fast-moving traffic not to be penalized by the slow-moving vehicles in the front. Overtaking assessment and assist helps in the overtaking task. It has two important tasks as follows:

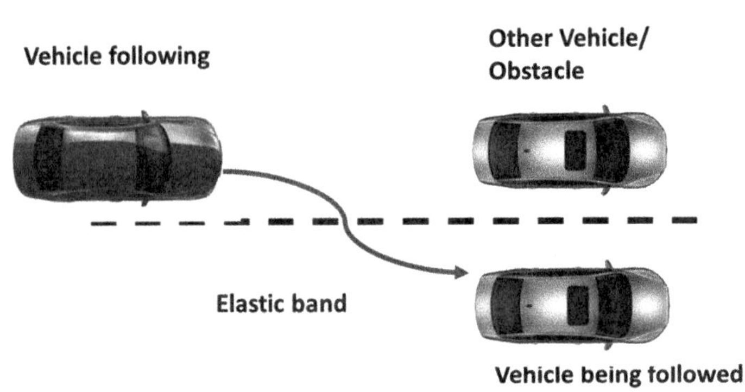

Fig. 10 Vehicle following

- Deciding the feasibility of the overtake attempt: The feasibility of the overtaking is based on the distance between the overtaking vehicles and the one being overtaken and the space in the overtaking lane
- Actual overtaking: This task requires moving the vehicle overtaking from the overtaking lane to the normal lane after overtaking as shown in Fig. 11.

One of the common problems faced by all drivers is to park the vehicle in an empty slot correctly. This problem can be solved by an automated parking system. The problem of parking resembles close to that of a robot motion planning. Some number of times, the vehicles goes back and forth before placing the vehicle perfectly in the empty slot of the parking area as shown in Fig. 12. The commonly used motion algorithms include rapidly exploring random trees, graph search, and many more [4].

Sensors Used in ADAS Technology

Various sensors are used in implementing the ADAS technology. Some of the main sensors are described in detail below:

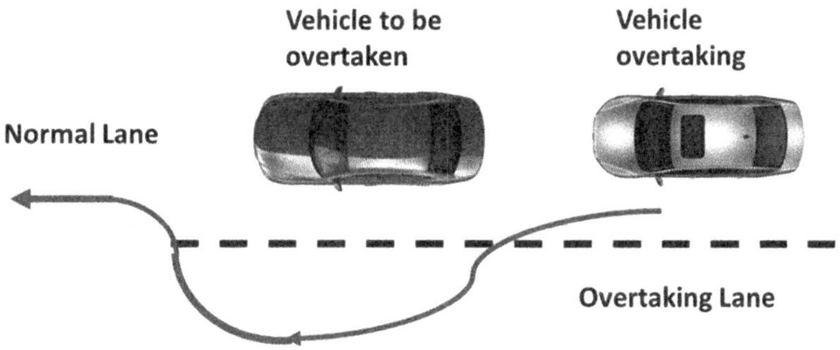

Fig. 11 Overtaking

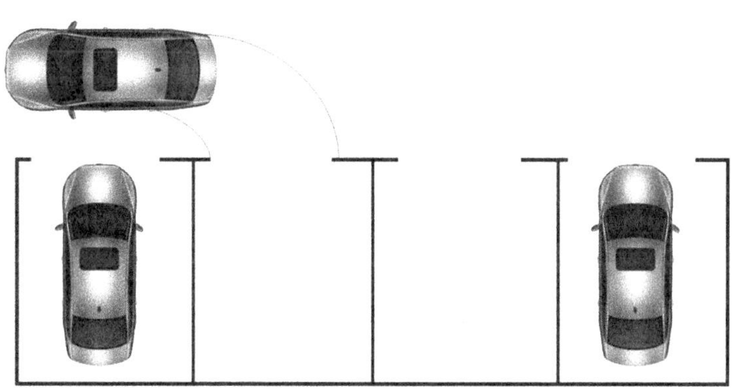

Fig. 12 Parking assistance with ADAS-based cars

Camera vision: It acts as the eyes of the ADAS technology. This vision is affected severely by weather conditions, low light, and glare. The image acquired is pre-processed using digital signal processors (DSPs). The camera sensors are typical used for applications such as traffic sign recognition, parking assistance, and lane departure warnings. Infrared cameras are used during nights which act as best night vision cameras.

Radio detection and ranging (RADAR): Compared to cameras, radars are robust in any weather and light conditions with good range. There are specific radars for long-range and short-range applications.

Light detection and ranging (LiDAR): The LiDAR measures the distances based on the light scattered w.r.t the laser irradiation that emits pulsed light. The main advantage of the LiDAR technology is that the light scattered from the object gives information distance as well as the nature of the object.

Ultrasound: These sensors work based on the principle of sound in determining the distance of the nearby objects and to warn the driver regarding the obstacles. Usually, a single car will be equipped with 12 ultrasonic sensors to completely cover the surrounding environment. The range of ultrasonic sensors are very less and is about 3 m.

The range and typical application of the sensors in the ADAS technology is tabulated in the Table 1.

Reliability Modelling

A vehicle consists of active and passive safety systems and ADAS features. The most important concept in ADAS is to maintain the safety by saving the lives. A model is demonstrated here to represent various ways on how the safety of the

Table 1 Various sensors used with typical applications

Sensors	Maximum range (m)	Typical applications
Short-range radar	20	Blind spot detection
Long-range radar	150	ACC (adaptive cruise control)
Camera vision	80	Pedestrian detection, sign recognition, parking assistance, lane departure warning
Infrared	120	Night vision
Ultrasound	3	Park assist
LiDAR (scanning)	120	Emergency braking, collision avoidance systems

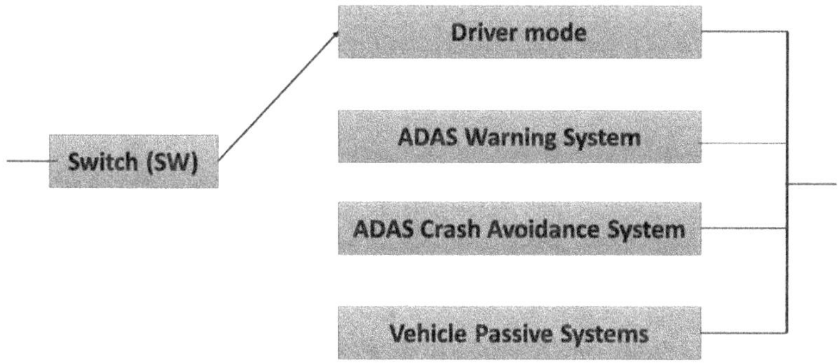

Fig. 13 RBD block diagram for a simple model

vehicle is maintained. The model represented is termed as reliability block diagram (RBD) and is shown in Fig. 13. There are four different pathways to maintain the reliability of the safety of the vehicle.

When the driver fails to operate the vehicle safely, the control systems alerts the ADAS warning system to provide the necessary warning signals. When the failure in ADAS warning systems result, it leads to the activation of ADAS crash avoidance systems. Lastly, the failure of ADAS crash avoidance systems switches on the passive safety systems in the vehicles such as airbags to active state.

Vehicular Communication in ADAS

Vehicular communication system plays a vital role in ADAS by notifying useful information through various channels like wireless communications and to implement the necessary actions especially regarding safe driving alerts. Safety is a key factor of an automotive. Every year, the increase in number of road accidents can be directly linked with increasing number of vehicles on the road. According to the WHO, globally at least 1.2 million deaths are happening every year due to road accidents. Traffic management is necessary to avoid possible accidents, to minimize carbon emissions, to improve fuel efficiency, for navigating the fastest route, to improve driver safety and comfort and to save resources. Several automobile manufacturers are doing research into the same to provide better future prospects for automobile industry. Vehicles could communicate with vehicles and other elements which substantially improvise the driving systems and safety prospects. Vehicle-2-vehicle (V2V), vehicle-2-infrastructure (V2I), vehicle-2-pedestrian (V2P), vehicle-2-cloud networks (V2N), and vehicle-2-everything (V2X) communication systems allow the vehicle to communicate to its surroundings with short-range wireless signals which is as depicted in Fig. 14.

Vehicular communication conveys to the driver about weather information, traffic and signals, road conditions, accidents, and any harmful activities around

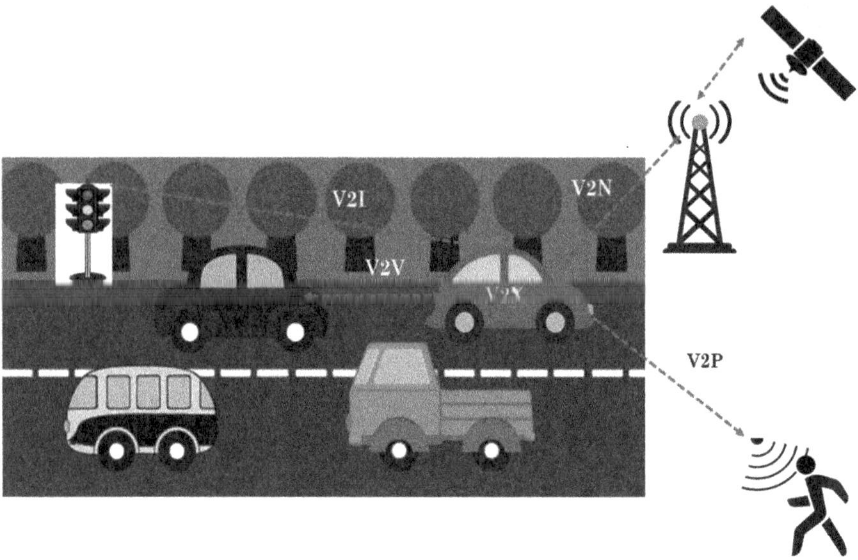

Fig. 14 Vehicular communication system in ADAS

vehicle. In autonomous vehicles, having the navigation system adds extra features when compared with traditional vehicles. One of the vehicular communication systems, V2X, has a smart working system which reports automatic payment for toll, parking slot identification, weather forecast, shortening the distance towards the destination, fuel station/restaurant notifications, etc. V2X communication has several components which integrate information that connect vehicles to everything (pedestrians, vehicles, roads, and cloud environments) aiming to reduce the number of road accidents and building an intelligent transport system (ITS) [5].

Certain features are added with ADAS by considering safety concerns like Emergency Electronic Brake Lights, Forward Collision Warning, Intersection Movement Assist, Left/ Right Turn Assist, Blind Spot/Lane Change Warning, Curve Speed Warning; by considering environment concerns like vehicle classification, traffic management, intersection analysis, destination studies, emission control, smart parking and by considering roadside mobility concerns like emergency communications and evacuation, cooperative adaptive cruise control, wireless inspection, connection protection, etc. Adding all these features with ADAS, vehicular communication will be a major advancement in the field of automotive by adopting technologies like Internet of Things (IoT), information and communication technologies (ICT), and artificial intelligence (AI), leading to safe driving environment [6].

Vehicular Communication Technologies Used in ADAS

Cellular systems, Bluetooth, microwave communication, millimeter wave communication, DSRC (dedicated short-range communications), ZigBee, WiMAX, and long-term evolution (LTE) are key wireless communication technologies used in vehicular ad hoc networks (VANETs) for effective vehicular communication. Various vehicular communication technologies used in ADAS are listed in Table 2, which could be used as wireless communication channels to connect the vehicle with everything around it depending on the applications.

Pathway Towards Nextgen: ADAS in Global Automotive OEMs

The advancements in terms of safe driving and automation carried in the form of ADAS had already arrived in OEMs globally. A leading Japanese car manufacturer introduced driving safety support system (DSSS) to its products Prius, Crown Majesta, Crown Royal, and Crown Athlete. DSSS have sensors and send messages to the driver about detected pedestrians and vehicles that were not visible to the driver's sight and alert the driver. The main benefits of DSSS are reducing the traffic accidents at intersections, increasing awareness, and reducing the driver's burden. DSSS features include pre-collision warning and braking, enhancing the road safety during night drives. DSSS uses V2V, V2I, and V2X communication systems. DSSS comes under the awareness driving phase which is the present trend in automotive sectors. Another car manufacturer hailing from Japan, Honda, thrives to demonstrate advanced V2P and V2V through DSRC communication technology. The ultimate aim of this R&D work ensures the reduction in collisions between vehicles and pedestrians. Next phase sensing and driving also make impact in the automotive sector about hazardous things on the road. In 2017 Daimler introduced autonomous driving with Mercedes E-class car using V2X communication. The sensing and driving principles had been applied to identify weather conditions and road accidents, intended to increase safety and reduce the stresses associated with driving. Few other automotive OEMs are involved in R&D of self driving cars. The third phase of advancement in ADAS comprises cooperative driving which enables V2V communications effectively. This phase operates multiple driver functions simultaneously. The German-based automotive, Volkswagen, adopted V2V communications for its vehicles from 2019, using public wireless LAN (pWLAN) with IEEE 802.11p standards. This communication enables to connect with vehicles and infrastructure which added some features like lane merger assistance, platooning, pedestrian safety, and parking reservation.

Further into the advancements in ADAS for next generation, the synchronized cooperative driving provides the features like overtaking assistance and intersection assistance. Most of the decisions would be made by the vehicle, but it requires the driver to supervise the actions. The automation is limited in this phase in terms of

Table 2 Vehicular communication technologies in ADAS

Vehicular communication technologies	Bandwidth	Purpose	Applications	Standards
DSRC (dedicated short-range communications)	5.95 GHz band	To provide high data transfers and low communication latency in a small communication area on vehicular networks	DSRC has a wide range of applications such as V2V safety messages, traffic information updates, toll fee collection, and several other applications	EN 12253 EN 12795 EN 12834 [ISO 15628] EN 13372 SAE J2735
WAVE (wireless access in vehicular environments)	5.9 GHz band	The communication between vehicles and vehicle to infrastructure with continuous and interoperable services	Avoid car accidents, parking the vehicle, identification of fuel stations and restaurants, and several other applications	IEEE 1609 2013 IEEE 1609-2010 IEEEP1609 IEEE 1609.11-2010 IEEE 1609.1 IEEE 802.1-2012
CALM (communications access for land mobiles or communications, air interface, long, and medium range)	5.9 GHz band	A set of wireless communication protocols and air interfaces for a variety of communication scenarios spanning multiple modes of communications and multiple methods of transmissions in ITS	Internet access, dynamic navigation, safety warnings, collision avoidance, and ad hoc networks linking multiple vehicles	ISO 29281 ISO 15628 ISO 24102
Infrared (IR)	Unlicensed band up to 100 Mbps and even more	Short-range communication	Vehicle information and communication system (VICS) in Japan and the truck tolling scheme in Germany	ASTM E1897 ASTM E 1933 ASTM E 1934
Bluetooth	Unlicensed band at 2.4 GHz over small distances of 10 m	Short-range communication with low power, low cost, low complexity, and robustness	Provide wireless connection between different portable or fixed electronic devices in the vehicle basically suited for communication, such as onboard communication devices	IEEE 802.15.1

ZigBee		20–250 Kbps which covers small distances of 100+ meters	Improving network services, multichannel operations, and resource management in ITS to support sensing, monitoring, and control applications with the lowest power consumption	The ZigBee sensor measures the distance between the vehicle in front of the car and will warn the vehicle in case there are any chance of collision. Emergency braking of vehicle moving in front. Supporting large number of vehicles on the road	IEEE 802.15.4 IEEE 1609
VLC (visible light communication)		Up to 500 Mbps	VLC uses lighting sources as transmitters and utilizes photodiodes as receivers for establishing communication	Environmental monitoring	IEEE 802.15.7
Cellular system	2G/3G	Up to 25 MHz	Cover an enormous geographical region which gives continuous communication to vehicles	GPS, short messages	Network standards
	WLAN/Wi-Fi (wireless fidelity)	2.4 GHz 54 Mbps covers at least 160 M	Efficient wireless access to empower intelligent transportation between vehicles and the infrastructure	Safety on the road, traffic flow, and comfort for passengers and drivers with expedited applications such as the Internet, network games, automatic electronic toll collection, drive-through payments, and digital map update	IEEE 802.11 IEEE 802.11a IEEE 802.11a
	LTE-4G	Up to 1700 MHz	Consumers would be able to access network- and cloud-based applications	Entertainment, infotainment, diagnostics, navigation, non-safety applications, such as traffic information transmission	IEEE 802.11p
	WiMAX (worldwide interoperability for microwave access)	Up to 20 MHz	To provide seamless wireless access over extensive distance	WiMAX can be effectively utilized for V2I or I2I communications, and it also provides long-range communications, which can be used for various applications	IEEE 802.16e
	5G—Millimeter WAVE	60–64 GHz	Autonomous driving, vehicle-2-vehicle (V2V), vehicle-2-infrastructure (V2I), vehicle-2-network (V2N), and vehicle-2-pedestrian (V2P) communications, enable new safety-sensitive applications V2V/V2I (V2X)	Alerts, speed recommendations, real-time updates Software updates, image sharing, automated platooning, situational awareness, and intersection management/safe intersections	IEEE 802.11ad

proper environmental conditions and complex situations which arise from the roadside. A French automotive OEM, Renault, developed communications between vehicles and roadside infrastructure such as toll booth, intersection area, school/college zones, and others. Sanef, a French motorway infrastructure, collaborated with Renault to develop features. Another example was set by the US-based General Motors division Cadillac – CTS Sedan was offered with V2V communication in 2017. They tie up with the Australian company Cohda Wireless for communication assistance and the Israeli firm, NXP Semiconductors, for developing road links for connecting with vehicles. Delphi automotive supplied antenna modules and provided data security.

The autonomous driving phase is envisioned with a fully autonomous vehicle. An autonomous vehicle would be able to make complex decisions on its own with all kinds of situations like environmental, infrastructure, and intersection conditions. The appropriate decisions are taken by the vehicle with no human intervention. The capability of reaching the destination can be fully controlled by vehicle without any concerns. One of the automotive OEMs, Land Rover, is currently working with an autonomous urban drive system under SAE level 4 standards of high automation. An autonomous car would be capable of communicating with cars and infrastructure around it (similar to V2X communication) which leads to the development of highly automated vehicles (HAVs) for long-term purpose.

Summary

This chapter discusses the ADAS technologies and its various aspects in improving the vehicle and community safety on roads. The main points discussed in this chapter on various ADAS features and the communication technologies are summarized in the following points:

- Detailed descriptions of various information-based and manipulation-based ADAS systems
- Various vehicular communication technologies used in ADAS
- Various sensors and their usage in ADAS-based systems
- Case studies and current trends in the implementation of ADAS

ADAS system have many positive impacts such as mitigation of exposure to risky conditions and improvement of driver behavior (e.g., reduced driving speed and speed variability, smaller lane deviations, faster reaction times, less harsh braking, and enhanced alertness) and eradication of driver errors. However, there are negative impacts also that are associated with the ADAS technology that include the following:

- Shifting of the driver's attention to road environment information that results in insufficient attention to the important driving tasks

- Frustration of the driver with the warning systems due to unnecessary frequent system warnings
- Frustration of the driver when certain elements of the driving tasks are taken over by the system in contrast to the driver's desire

References

1. S. Singh, *Critical Reasons for Crashes Investigated in the National Motor Vehicle Crash Causation Survey* (NHTSA, Washington, DC, 2015) Report No. DOT HS 812 115
2. R. Spicer, A. Vahabaghaie, G. Bahouth, L. Drees, R. Martinez von Bülow, P. Baur, Field effectiveness evaluation of advanced driver assistance systems. Traffic Inj Prev. **19**(sup2), S91–S95 (2018)
3. R. Kala, *On-Road Intelligent Vehicles* (Elsevier, Amsterdam, 2016)
4. L. Han, Q.H. Do, S. Mita, Unified path planner for parking an autonomous vehicle based on RRT, in *2011 IEEE International Conference on Robotics and Automation*, (Shanghai, 2011), pp. 5622–5627
5. F. Arena, G. Pau, An overview of vehicular communications. Future Internet **11**(2), 27 (2019)
6. A. Paul, N. Chilamkurti, A. Daniel, S. Rho, *Intelligent Vehicular Networks and Communications Fundamentals, Architectures and Solutions* (Elsevier, Amsterdam, 2017)

Analysis of IoT-Enabled Intelligent Detection and Prevention System for Drunken and Juvenile Drive Classification

D. Ruth Anita Shirley, V. Kamatchi Sundari, T. Blesslin Sheeba, and S. Sheeba Rani

Introduction

A proper connection of the Internet through wireless technologies and an embedded sensor is all that is needed to make an object "smart" in today's world. The term "smart" was associated with phone since 1992 and was eventually introduced in the market as "smartphone" in 1995. In 1999, smart watches began making their way into the market, but it was only in 2012 were they recognized for the immense potential that they presented with their technological advancement in using multiple applications from the Android phones. Other wearable devices have also been introduced in the past decade, offering unprecedented opportunities for personalized services, real-time health monitoring, and human behavior modeling.

When one says "IoT," it brings to mind watches, phones, and other similar devices. But, the scenario has changed now, and the automobile industry has progressed greatly, through experimentation with IoT and how it can be linked with automobiles [1], making it more efficient. The advancement of technology which started with embedding GPS (global positioning system) for navigation has now led to analyzing and gathering data about the vehicle with the help of sensors. It is predicted that by 2030 every car will be automatically connected to the insurance

D. R. A. Shirley (✉)
Department of ECE, Sri Krishna College of Engineering and Technology, Coimbatore, India

V. K. Sundari
Department of ECE, Prince Shri Venkateshwara Padmavathy Engineering College, Chennai, India

T. B. Sheeba
Department of ECE, RMK College of Engineering, Chennai, India

S. S. Rani
Department of EEE, Sri Krishna College of Engineering and Technology, Coimbatore, India

M. Kathiresh, R. Neelaveni (eds.), *Automotive Embedded Systems*, EAI/Springer Innovations in Communication and Computing,
https://doi.org/10.1007/978-3-030-59897-6_10

carriers, service companies, manufacturers, and even the drivers, across the globe. In fact, almost all the cars that are built today have over 400 sensors on them to gather data regarding car wear and tear, GPS positioning, speed, driver actions, tire pressure, steering wheel movement, and much more. Despite the advancement in technology, there is still human error in driving the car, leading to fatal accidents.

A survey by the World Health Organization reports that road traffic injuries hold the eighth place in causing death, accounting for 2.5% globally. It has been observed that road traffic injuries are also the leading cause of death in people aging between 5 and 29. Due to progress in medicine, infectious diseases are now in check with control and preventive measures. This has pushed road traffic injuries to reach the eighth place in causing deaths compared to the ninth position it was previously holding, as shown in Fig. 1. It has been recorded in the WHO report [2] that there is a direct relation between the income of a country and the risks of road accidents leading to death. In fact low-income countries like India and China face more deaths (about 13%) than high-income countries like Switzerland and Greenland which account for about 1% death.

Smart vehicles running on smart brains through embedded sensors, wireless communication, and IoT promise to enhance security, safety, and comfort of the driver, the occupants, as well as the pedestrians. Accidents/collision avoidance, obstacle detection, lane departure, automatic brake control, air bags, theft notification, and automobile tracking are some of the innovations in technology that have been implemented or are in the phase of development to ensure safety of the travel

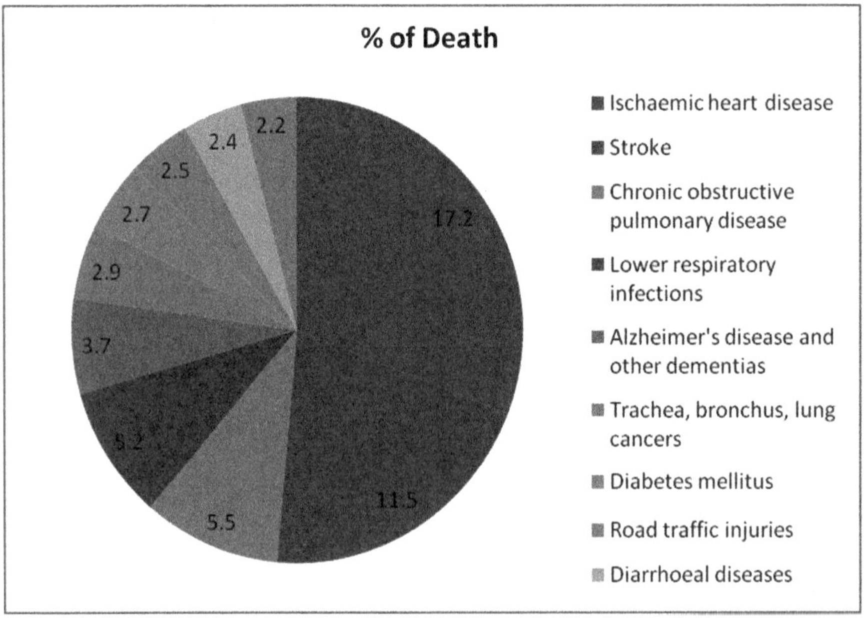

Fig. 1 Percentage of death recorded by the WHO during the year 2016

in automobiles. This has paved way to the beginning of the era of intelligent vehicles. At the same time, drivers have become irresponsible, ignorant, and careless in driving. There is continuous report of reckless or rash driving and driving under intoxicated state/drunken driving which have led to accidents, injuries, and even life loss of drivers and innocent pedestrians.

Based on the survey by the National Crime Records Bureau, it has been identified that India is one of the top 10 countries in the world with high road accidents occurring every minute. The initiative by India to curb liquor shops in some cities and near highways has proven to be fruitful in bringing the number of accidents down by 36.5%. However, the resultant figure is still quite high at 12,018 road accidents in which 4188 persons have died during the year 2018 alone, as shown in Fig. 2.

Survey reports that over 70% of the pedestrians are affected by rash driving of drunken drivers. To put an end to this case of drunken driving, a graphene sensor will be fitted on the steering wheel to control the start of the car engine. In 2019, researchers from Uttarakhand Residential University and RI Instruments and Innovations had experimented and designed a graphene sensor which is said to raise

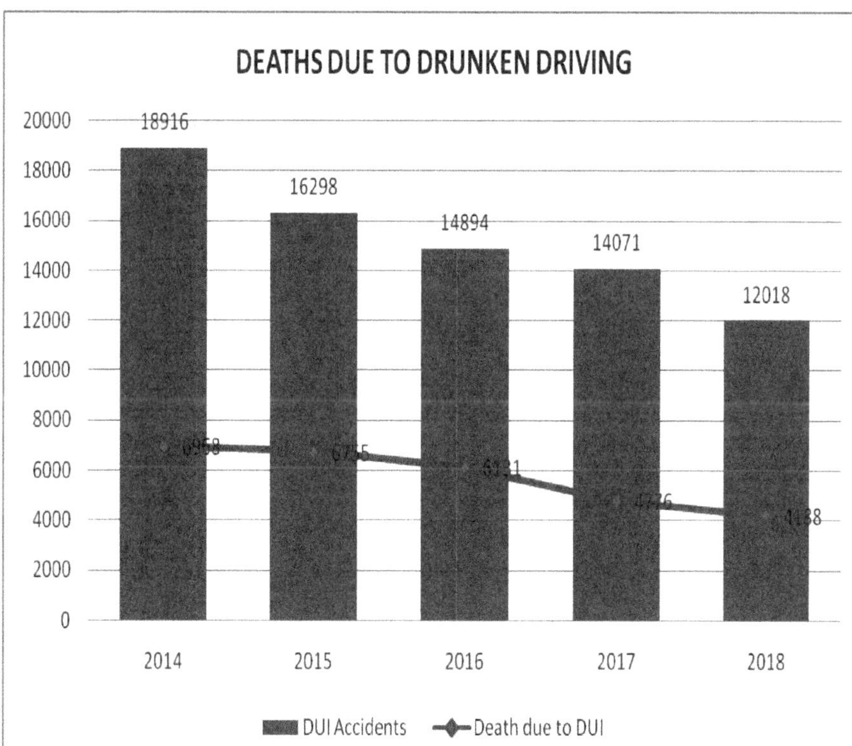

Fig. 2 Deaths due to drunken driving against DUI accidents caused during the years 2014, 2015, 2016, 2017, and 2018

an alarm when a drunk driver blows on it. The ethanol level is diagnosed by the sensor, and if the threshold has been crossed, the sensor will stop the driver from operating the car.

With an alarming increase in the rate of juvenile driving accidents, we have also proposed a technology to scan the fingerprint of the driver in order to start the engine. The database of the drivers can be saved in the cloud and checked at a regular interval. When the conditions of the graphene sensor and the fingerprint scanner are satisfied, the engine can be started by the driver.

Road safety through smart automation has received increased interest recently, and a number of innovative ideas and solutions have been implemented over the past decade. This work aims to curb the usage of automobiles by drunken drivers and juvenile drivers, thereby reducing the number of accidents on the road and ensuring the safety of pedestrians and other drivers.

In this work, the existing methods for drunken driving prevention are analyzed along with a record of juvenile driving accidents that have occurred in the recent years.

The book chapter proposed is organized as follows:

- Background and Pertinent Work: in this section the existing literature on driver identification and alcohol observation using biometric scanning technologies are summarized.
- Methodology proposed: the proposed methodology and the working of the methodology are outlined in this section.
- Results and discussion: the results of the proposed work are recorded on observation, and the engine is allowed to start based on the analysis of the result.
- Conclusion and future scope and discussion: concluding remarks and possible future enhancement of the proposed work are mentioned at the end.

Background and Pertinent Work

Between 2006 and 2016, the number of accidents involving intoxication and drunk driving was very high, with an average of 10,000 deaths caused by such accidents [3]. Survey indicates that 1.3% of car accidents are caused due to medical emergencies like diabetic reactions, blackouts, and seizures [4]. Rathore and Gau [5] designed a wearable biometric sensor named as the Automotive Healthcare and Safety System which was used to keep track of the user's body temperature and heart rate. It also incorporated a breath analyzer device which could determine if the driver was drunk or intoxicated. To check if the driver has any medical emergency, infrared cameras were placed on the dashboard to perceive facial symptoms and pupil dilation. An ECG monitor was also placed on the seat belt to detect weight and movement. Based on the analysis of these biometric sensors, emergency responders were notified, when any discrepancy was identified apart from the normal level of detection.

Over the years, a number of techniques and technologies have come into the market to prevent or reduce the number of drunken driving accidents by automobiles. In the paper by Lee et al. [6], an alcohol detector comprising of liquid crystal display (LCD) circuitry, LM358 operational amplifier (Op-Amp), and alternating current (AC) power supply is built. This alcohol detector will sense the alcohol content present in the surrounding air, and when the alcohol content in the air crosses a predefined threshold value, it will display a warning message.

However, since the device operated on AC power, the person to be tested for drunken driver had to sit close to the AC power outlet. Another drawback was that the Op-Amp (LM358) which was used as a comparator had "5" as the preset threshold value and did not raise any alarm of warning when the alcohol level crossed the threshold. Hence, this constrain required two participants to test and note the alcohol level when using the alcohol detector, preventing it from being positively used to test real-time drunken drivers.

Another alcohol detection system which used a cell phone to alert the driver was proposed by James and John [7]. This system comprised of a LM358 module and a GSM technology as the major components. Since it was dependent on the GSM technology unit, it was a huge step forward from the breath analysis used till then. The GSM technology made use of a GSM module that replaced the traditional alarm system while making use of LM358 Op-Amp.

To distinguish the text message from this GSM module, a unique ringtone was also defined for the alerts on the cellular phone. The drawback of this system is that when the driver is not able to access his phone or when the phone is out of network coverage, this system proves to be inefficient and insufficient. To give flexibility in changing the concentration of blood alcohol, LM358 Op-Amp cannot be used.

Based on this observation, a PIC16F877A microcontroller alcohol detection system was built by Phani et al. [8]. The PIC16F877A microcontroller gave the users the flexibility of changing the threshold of alcohol to be detected based on body chemistry. Moreover, the use of a microcontroller also opened up opportunities for adding more features to the alcohol detection system. The drawback of this proposed system [8] is the inability to implement it in real time as it requires AC power supply instead of a DC power supply to operate to its optimum degree of accuracy.

Gestion [9] also proposed a similar system to monitor the health of the driver through machine learning algorithm, skeleton tracking with gesture recognition, and synthetic depth data. The proposed system could monitor the health conditions of the driver and detect health problems. A breath analyzer was embedded in the steering wheel of the automobile to keep track of the alcohol level of the driver by sober steering [10]. The proposed system will raise an alarm when the predefined threshold value of alcohol was surpassed by the driver. Taking this proposal one step forward, Chen and She [11] proposed a device that was created to prove sobriety of the driver to start the car.

According to the report by the World Health Organization (WHO) on road accidents in India (2018) [12], 32,438 accidents have occurred due to drunken driving, in the observed 16 states in India, of which 11,660 deaths have been incurred. The resulting death rate due to DUI is almost one-third of the accident rate. A survey of

Table 1 Chart of persons killed in individual states from road accidents during the years 2016, 2017, and 2018

State	No. of persons killed			No. of accidents		
	2016	2017	2018	2016	2017	2018
Jammu and Kashmir	21	0	5	47	1	20
Himachal Pradesh	17	89	70	72	214	322
Punjab	197	95	85	317	129	112
Uttarakhand	36	30	16	40	56	20
Haryana	165	86	251	529	180	474
Rajasthan	372	142	52	673	421	146
Uttar Pradesh	2716	1687	1824	4633	3336	3595
Gujarat	23	18	15	64	65	106
Madhya Pradesh	457	217	145	3083	1049	893
Jharkhand	408	430	328	543	801	517
Maharashtra	58	293	42	226	863	188
Karnataka	94	23	22	396	169	139
Telangana	55	34	24	202	163	182
Andhra Pradesh	45	156	85	128	2064	1345
Tamil Nadu	65	403	225	531	1833	1128
Kerala	7	7	25	133	133	157

the accidents and deaths in individual states indicates that Uttar Pradesh is the state with the highest number of accidents due to DUI and subsequently a high number of deaths at 6227 during the years 2016–2018, as shown in Table 1.

On analyzing the driver's biometrics, one can detect the identity of the driver. Furthermore, this can be enhanced to detect if the driver is under alcoholic influence and action can be taken accordingly. Gutmann et al. [12] have proposed a method to determine the stress level of the driver using neural networks by means of taking the hip and wrist movements, blood oxygen level, electrodermal activity, body temperature, inter-heartbeat interval, heart rate interval, breathing pattern, heart rhythm, and brain waves. Based on the results, the driver's psychological fitness was determined to grant him access to the automobile. Though it is easy to determine the psychological state of the driver with the help of biometrics, it will be a hindrance to the driver who has to attach multiple sensors on his body to check the ECG signal.

To overcome this drawback, Lee et al. [13] proposed a method to observe the ECG signal of the driver by placing the electrodes on the steering wheel to measure RR interval, heart rate variability, QT interval, ST segment, waveform, and heart rate when the driver places his hands on the wheel. In fact, it was found that this method of monitoring ECG had better clarity in results when compared with the traditional method of taking ECG used by Gutmann et al. [12].

Distractions inside the vehicle due to talking, mobile phones, smoking, eating and route guidance systems were studied by Young and Regan [14]. He proposed a system which could measure the distractions caused by monitoring the eye movement and glance action of the eye. Dehzangi et al. [15] have proposed a system that uses body sensor network [16] to collect biometric data and correlate with road conditions to monitor driver distractions. At the same time, automotive sensors are fit to monitor driver behavior like speed control and aggressiveness while driving. This information is collected from four drivers driving during different traffic conditions and is correlated with the road conditions. The drawback with this mechanism is that it will be restricting and invasive to the driver who might get distracted with the sensors and BSN [17].

Hence a less invasive device was built by Harman International Industries to monitor the pupil dilation of the driver with the help of a camera [18]. When the device detects a rise in cognitive load, it will alert the driver about the distraction and further enable the "silent" mode of the driver's phone.

Another common cause of accidents is drowsiness of the drivers due to sleep deprivation or sleeping disorder. To spot signs of drowsiness, Panasonic Corporation made use of artificial intelligence algorithms [19] and ensured that the driver was well awake during his time driving. Facial expression, body heat loss, and eye blinking rate were observed with sensors and camera. The data gathered during this process is further analyzed, and when signs of drowsiness are observed, the system will automatically adjust the AC settings and the volume of stereo to keep the driver awake.

A similar method was devised by Omron [20], in which an infrared camera was placed on the dashboard of the automobile. This infrared camera will monitor the head movements, blinking rate, gestures, and eye movement. If there is any indication of drowsiness, the car will automatically come to a halt. The use of infrared cameras was found to be inefficient and space-consuming. This led to the development of smart glasses [21] that could capture data directly from the eye. In line with this, Vigo manufacturers developed a headset that was capable of detecting if the driver was drowsy [22].

To ensure that the vehicle is safe and secure, biometrics-based authentication goes a long way in fighting theft. It also replaces the traditional aspect of carrying a key to unlock the automobile, which is also prone to be lost or stolen. Biometrics on the other hand cannot be stolen or replicated. Fingerprint, voice, and face are three common ways of biometric recognition techniques [23].

Mercedes, BMW, and Lexus [24] have already implemented keyless access for facial recognition, and other similar companies have similar progress. Ishak et al. [25] developed a system which uses face recognition to identify the driver and allows him to start the car, if his face features match a predefined database. During the night, it will be difficult to identify the face of the user, and so illumination conditions were implemented to make it work efficiently. A 480 × 640 pixel camera was used to capture the face of the driver. Classical neural network and fast neural network [26] have been used to detect the face of the driver which is further analyzed using linear discriminant analysis and principal component analysis [27].

Fingerprint technology has had wide success in mobile phones as a part of high-security applications. Incorporating this technology in automobiles will also be useful when using keyless access option. Zhu and Chen [28] in 2011 had come up with an idea to check the fingerprint of the driver before starting the vehicle. This has been the base for fingerprint usage in automobile, and many advanced technologies have been developed for recognizing fingerprints.

Sushmitha et al. [28] used a libfprint open source fingerprint processing library to verify the fingerprint of the driver. On verification, the device will give the driver access to open and start the automobile. A FPM20A fingerprint module was used by Folorunso et al. [29] to identify the fingerprint and allow/deny access to the driver.

Methodology Proposed

Capacitive Scanners

There are many fingerprint scanners available in the market, and the capacitive scanner is one of the most commonly used scanner because of its security and thin plate structure. The core component of the capacitive scanner is the capacitor. The capacitor circuits collect information about the fingerprint that is placed on the scanner instead of creating an image of the fingerprint. This feature helps detect the fingerprint easily and without any delay when compared with the conventional way of optical fingerprint sensor.

How Does the Capacitive Scanner Work?

The capacitive scanner is built with the basic component as capacitors. Capacitors are charge-storing and by connecting them to conductive plates, the scanner will enable us to keep track of the fingerprint details. An array of capacitors will be placed underneath the scanner to gather details of the fingerprint, as shown in Fig. 3. When the ridge of the finger is placed on the conductive plates of the capacitor, the charge stored by the capacitor will vary. On the other hand, the charge will remain unchanged where there is an air gap. The changes observed in the capacitors are tracked with the help of an Op-Amp integrated circuit, and this data is saved by an A-D converter.

The data captured through the capacitive scanner are segregated for unique fingerprint characteristics. Fingerprints of the users are saved in this manner to compare against the driver prints before starting the engine. Depending on the count of capacitors placed in a pixel array, the resolution capacity of the capacitive sensors is determined. To enhance the performance of the recognition system, a large number of capacitors are used.

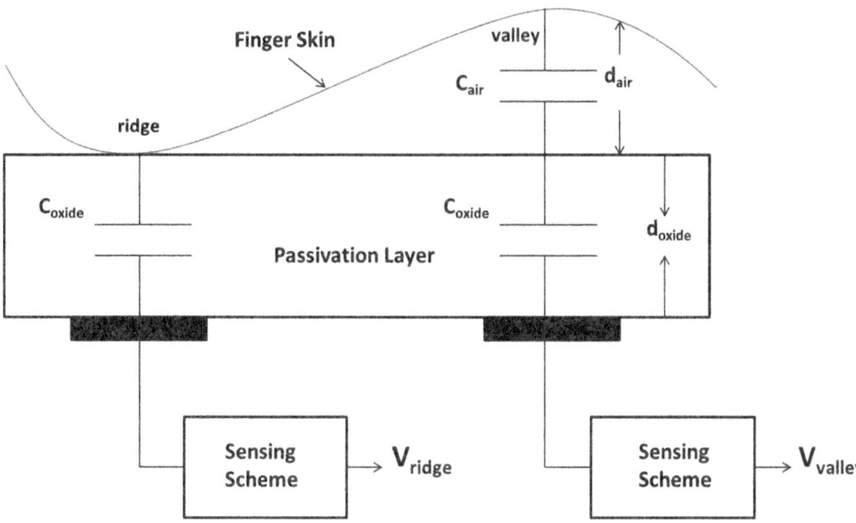

Fig. 3 Working of the capacitive scanner

Analyzing the Captured Data

The digital data that is captured is further analyzed to match its unique and distinctive attributes with that of the database. The capacitive scanners are more robust than the optical sensor, and its sleek design will give the builders the choice of installing it anywhere with comfort. The proposed design is focused on placing the capacitive scanner along the rims of the wheel for extracting the attributes on a regular interval, ensuring that the driver on the wheels is not replaced by anyone. Moreover, the best aspect of using this scanner is that it cannot be fooled with prosthetics as it will reflect in the charge stored by the capacitor.

When a large number of such capacitors are placed in a single scanner, it will ensure a solid database of valleys and ridges of a fingerprint. When the number of capacitors used increases, the quality of resolution will also increase, enhancing the level of security.

ARM Cortex-M Processors

There has been phenomenal progress in the development of the Cortex-M Processor family in the last 2 years. ARM Cortex-M processors provides a suitable platform for many applications including IoT. Security, memory space and addressing modes, high performance, high code density, and low power are some of the requirement for IoT applications, and the ARM Cortex-M series matches the needs of our design.

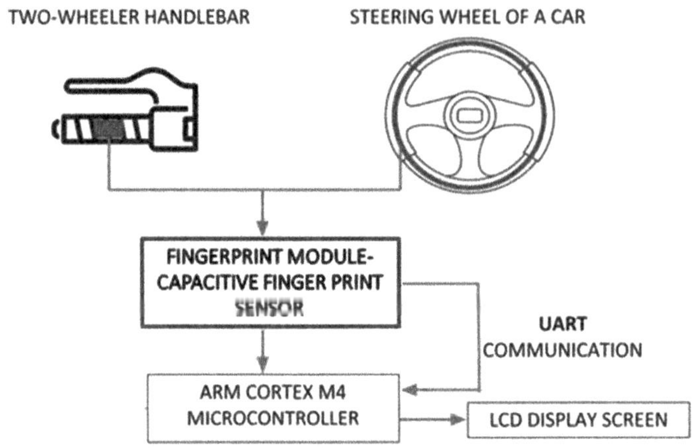

Fig. 4 Fingerprint module to capture the fingerprint from the driver and verify it against a predefined database

The capacitive sensor is placed along the rims of the steering wheel or the handlebar of the two-wheeler, as shown in Fig. 4.

Fingerprint Database

A predefined database of authorized users is stored into the database of the microcontroller to compare and allow the driver to start the automobile. However, the amount of information that can be stored within the microcontroller is limited. To overcome this limitation, the automobile is linked to the cloud storage through IoT, and the fingerprint scanned is sent to cloud for verification.

The proposed system is designed such that the cloud storage has a complete database of all the UIDAI users in India. Accordingly, the database will consist of fingerprint details of all users against which the detected fingerprint is matched. If the fingerprint belongs to that of a juvenile driver, an instant message will be sent, notifying the parent of the juvenile driver about the incident along with a message to the police station at the locality of the residence of the juvenile driver (obtained through the UIDAI database). However, since we do not have access to the UIDAI database, the system is tested for predefined users. In the future, this could be brought to the notice of UIDAI authorities to further implement the linking of UIDAI number to the automobile field.

In the past 3 years, a total of 30,007 accidents have occurred due to the juvenile drivers, of which 23,607 drivers were male and 6400 drivers were female (Fig. 5).

Considering this drastic accidents count, it is essential that a system is devised to curb such accidents at the root, and the proposed methodology is an efficient way of doing that.

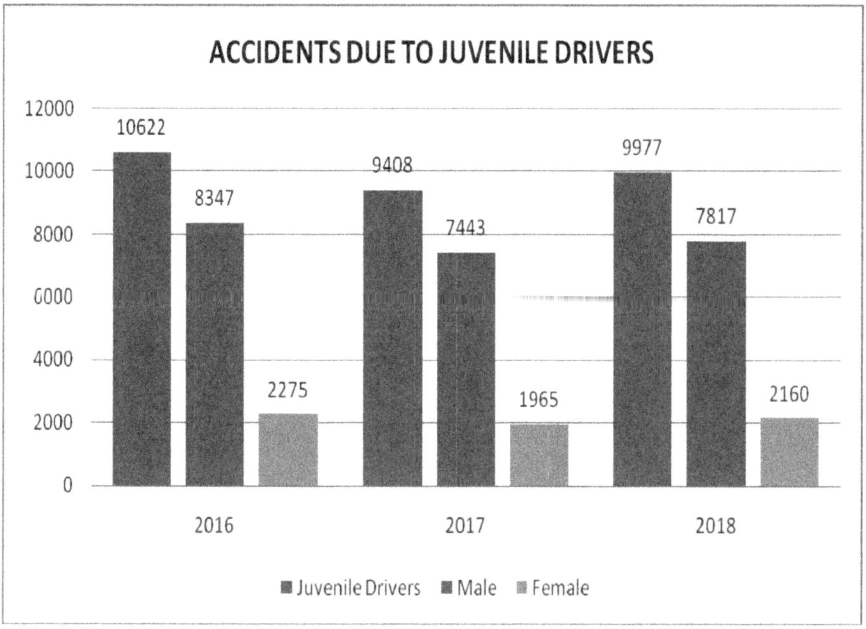

Fig. 5 Accidents due to juvenile drivers during 2016, 2017, and 2018

Drunken Drive Sensing

Driving under the influence (DUI) or drunk and driving is a serious offence and is punishable by law. In India, drunk and driving is one of the major reasons behind accidents on the road. According to the World Health Organization report, about 54.1% of people who drove under the influence of alcohol perished, while about 70% of the pedestrians were affected by the same. There are many mechanisms to detect the alcohol level in the person which mostly requires a blood test of the driver in question. To overcome this constraint, we propose the usage of a graphene sensor placed at the center of the wheel/below the calibrator of the handlebar to detect the alcohol level in the breath of the person driving the automobile (Fig. 6).

Alcohol Analysis

When a person consumes alcohol, it can be detected in his breath. Alcohol which enters the person is absorbed from the mouth and reaches the bloodstream. The alcohol content in the body will not be altered by any chemical reaction within the body, nor will it be digested. However, as the blood circulates through the lungs, some amount of alcohol will get evaporated through the alveoli and exhaled from the person. Hence a test of the breath of the driver, using a breath analyzer, can help detect the presence of alcohol.

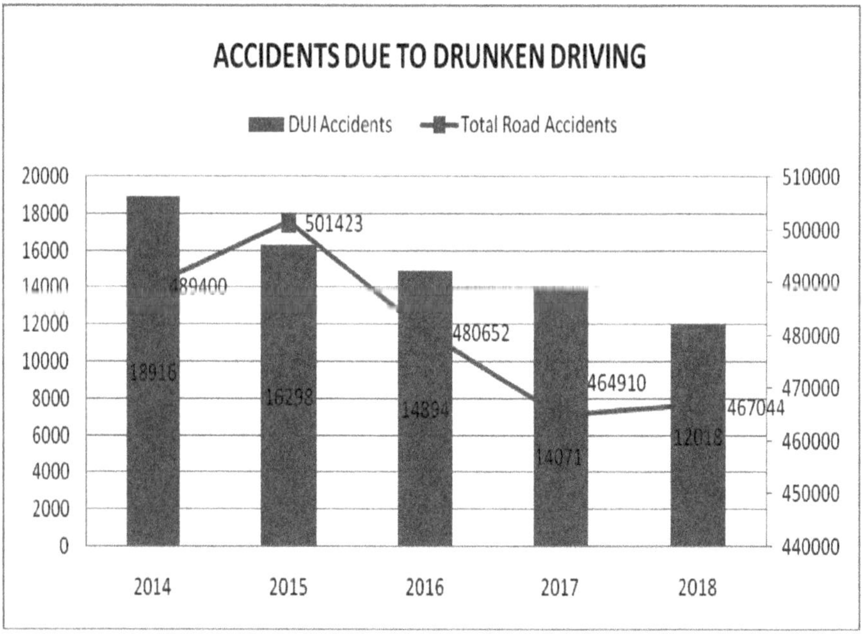

Fig. 6 Accidents caused due to drunken driving against DUI accidents caused during the years 2014, 2015, 2016, 2017, and 2018

When a drunken driver is halted to be checked, it is easier to test the breath of the driver for alcohol content than draw blood from him to spot for alcohol. This not only saves time but also helps quicken the process. The blood alcohol content (BAC) measured by the breath analyzer will indicate if the person can or cannot ride the automobile. 2100:1 is the ratio of breath alcohol to blood alcohol. This indicates that the measure of alcohol in 2100 ml of breath is equivalent to alcohol present in 1 ml of blood. In India, a person who has more than 30 mg of alcohol in 100 ml of blood (i.e., 0.03%) is considered unfit to drive an automobile.

Graphene Sensor

When a person is in under the influence of alcohol, there is an increase in the alcohol level in every breath he exhales. This increase in alcohol level is identified by the graphene sensor. Graphene is used in sensing applications for its advantageous properties like optical transmittance, charge carrier property, high flexibility, high thermal conductivity, high electrical conductivity, large surface area, and composition ease. Graphene has high adsorption capacity due to its ability to adhere to sheets with ease, which is also the reason why it is used as a common sensing device.

Carboxylated graphene (CG) and alcohol oxidase (AOX) ultrathin films which are used to detect alcohol content in the air can be easily fabricated and are economical and eco-friendly. When exposed to ethanol, they show highly sensitive amperometric response and are also designed with the ability to respond to ethanol in alcoholic beverages.

Tests on the graphene sensor show excellent sensitivity to alcohol level, enabling it to be an excellent detector of high sensitivity ranging from 920% in water medium. Unlike other composites, conductive filler material is not required for graphene because of its good conductive property. Along with the bacterial cellulose nanofibers, it serves as an excellent sensor to detect alcohol level in the breath of the driver. Graphene has the ability to catalyze the formation of acetic acid when ethanol comes in contact with air (oxygen). If a driver is drunk, the alcohol (ethanol) present in his breath will turn into acetic acid when it comes in contact with oxygen in the air. In turn, the acetic acid will be immediately identified by the sensor, signaling that the driver is intoxicated or drunk. The sensor will further analyze the amount of alcohol present in driver and will send the "Not Permitted" signal to the ARM controller, which in turn will disable the automobile from starting.

Figure 7 shows the graphene sensor placed along the rims of the four-wheeler and on the handlebar of the two-wheeler. In order to start the engine, the driver has to blow on the graphene sensor, and only if no alcohol trace is present in the breath of the driver will the system give access to starting the automobile.

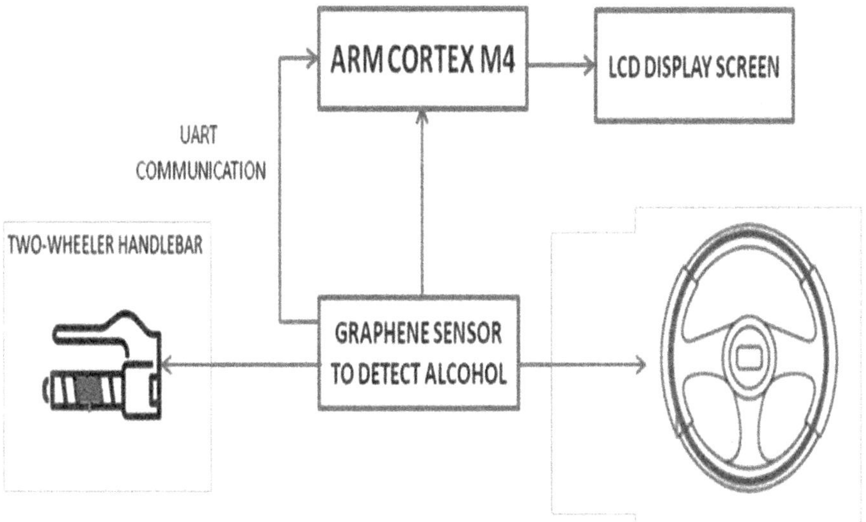

Fig. 7 Graphene sensor module to detect alcohol in the breath of the driver

Results and Discussion

Implementing the drunken driver and juvenile driver detection system requires five major components, namely, the fingerprint module capacitive fingerprint sensor, the arm microcontroller, the graphene sensor module, the LCD display, and a 120 rpm motor.

The proposed system was implemented in a two-wheeler to check if the system responded to the driver fingerprint and breathe analysis. Initially the fingerprints are loaded into the system before checking for its operation so that when the driver places his fingerprint on the capacitive scanner, the scanner will indicate if the driver is permitted to drive the vehicle. It will check the driver's fingerprint against the pre-collected and saved database On checking, if it is identified that the person accessing the vehicle is a juvenile, then the parent of the person is immediately alerted through SMS.

The driver who wishes to start the engine should first scan his fingerprint in the capacitive fingerprint sensor. When the driver is approved, he will be indicated to blow on the graphene sensor. If the alcohol content is within the predefined threshold range, the engine of the vehicle will start. When the finger is placed on the fingerprint sensor, the capacitive scanner (Fig. 8) will read the fingerprint and perform one of the following cases:

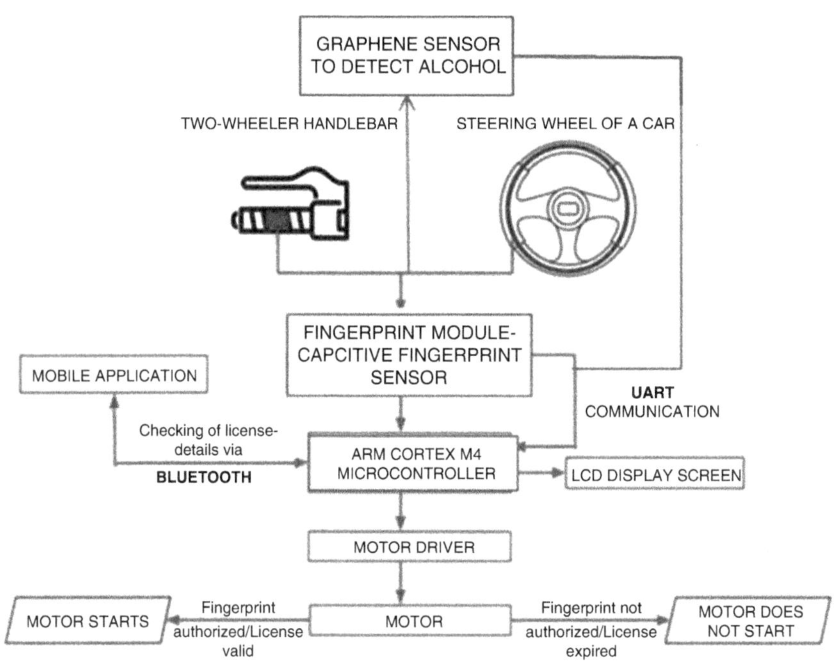

Fig. 8 Proposed block diagram to verify fingerprint and test for alcohol in the driver's breath

Case 3

If a new fingerprint which is not present in the database is detected, the motor does not run.

Figure 9 shows the LCD display indicating "Fingerprint Approved" to start the motor. Similarly, the four-wheelers can also incorporate this technology along the rims of the steering wheel (Fig. 10).

Conclusion

The proposed research work was successfully implemented in a two-wheeler automobile. The fingerprint of a predefined database was initially fed into the microcontroller. The driver who is to start the automobile was asked to verify his fingerprint. On verification, the driver was further requested to blow on the graphene sensor placed on the handlebar casing. Only if the alcohol level of the driver is below the predefined threshold value of 0.03%, the driver will be allowed to start the ignition of the automobile. This same methodology can be further incorporated in four-wheelers to prevent drunk drive accidents on the road. The proposed system has the following advantages:

- Prevents juvenile drivers from starting the automobile
- Prevents unknown users from using the automobile
- Prevents drunken drivers from driving the automobile

Future Scope

The proposed system has been implemented by feeding a predefined set of data for a group of drivers. However, if the same system could be synced with the UIDAI website, it can trace any driver using the vehicle from any corner of the country. The UIDAI is the government agency in India which provides a unique identity in the name of AADHAAR to the residents of India. A typical AADHAAR card will hold details which include name, date of birth, and address of an individual.

Further, the iris and fingerprints of the individual are recorded during the time of registration for the UIDAI. Hence, when the proposed system is interfaced with the UIDAI numbers, it will give the device access to find the age of the driver. If the driver is juvenile, he/she will be prevented from starting the automobile. Similarly when an accident is incurred, the police can trace the user who drives the automobile preventing hit-and-run cases from occurring.

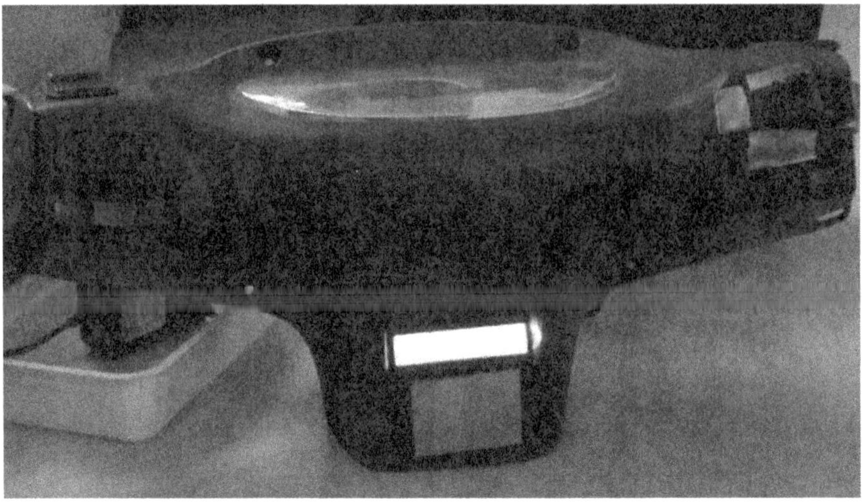

Fig. 9 Prototype model of the proposed system for fingerprint scanning and display of result

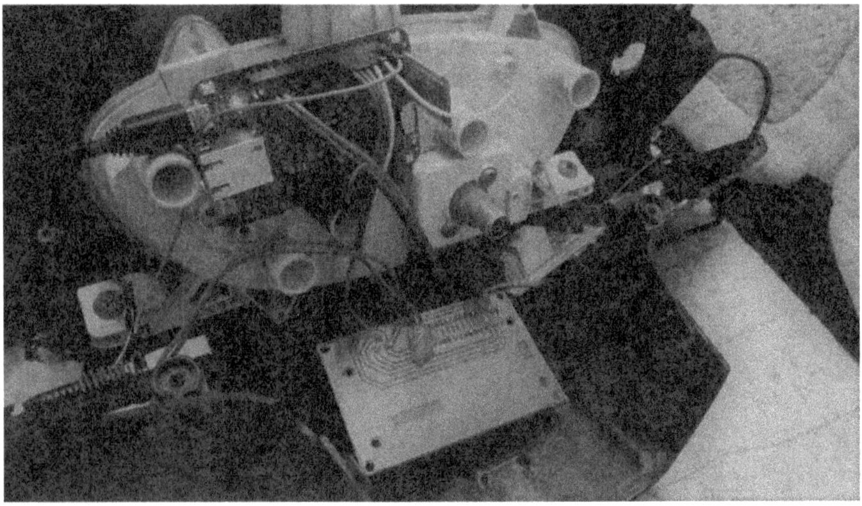

Fig. 10 Prototype model of the proposed system for fingerprint scanning and display of result

Case 1

If the detected fingerprint is a match with the database stored, the LCD displays "Fingerprint approved," and the motor operates.

Case 2

If the detected fingerprint is a match but is matched with a juvenile driver, the LCD displays "Parent Alerted," and the motor does not run.

References

1. R. Van Der Meulen, J. Rivera, Connected cars will form a major element of the internet of things. 2015. http://www.gartner.com/newsroom/id/2970017. Accessed 21 Apr 2017
2. Global status report on road safety 2018- WHO (2019). https://www.who.int/violence_injury_prevention/road_safety_status/2018/English-Summary-GSRRS2018.pdf
3. NHTSA, Drunk driving (2017). https://www.nhtsa.gov/riskydriving/drunk-driving
4. NHTSA, Contribution of medical conditions to passenger vehicle crashes (2009). https://crashstats.nhtsa.dot.gov/Api/Public/ViewPublication/811219
5. R. Rathore, C. Gau, Integrating biometric sensors into automotive internet of things, in International Conference on Cloud Computing and Internet of Things (CCIOT) (2014), pp. 178–218
6. N. James, T.P. John, Alcohol detection system. IJRCCT **3**(1), 59–64 (2014)
7. S.A. Phani et al., Liquor detection through automatic motor locking system: in built (LDAMLS). Int. J. Comput. Eng. Res. **4**(7), 2250–3005 (2014)
8. World Health Organisation, Global status report on road safety 2018, https://morth.nic.in/
9. Gestigon, The future of mobility is not about cars. http://www.gestigon.com/automotive-industry/
10. The power to stop drunk driving is now in the palm of your hand. http://sobersteering.com/how-it-works/
11. R. Chen, M. She, J. Wang, X. Sun, L. Kong, Driver verification based on handgrip recognition on steering wheel, in *2011 IEEE International Conference on Systems, Man, and Cybernetics (SMC)* (IEEE, New York, 2011), pp. 1645–1651
12. M. Gutmann, P. Grausberg, K. Kyamakya, Detecting human driver's physiological stress and emotions using sophisticated one-person cockpit vehicle simulator, in *Information Technologies in Innovation Business Conference (ITIB)* (IEEE, New York, 2015), pp. 15–18
13. H.B. Lee, J.M. Choi, J.S. Kim, Y.S. Kim, H.J. Baek, M.S. Ryu, R.H. Sohn, K.S. Park, Nonintrusive biosignal measurement system in a vehicle, in *29th Annual International Conference of the IEEE Engineering in Medicine and Biology Society, EMBS 2007* (IEEE, New York, 2007), pp. 2303–2306
14. K. Young, M. Regan, M. Hammer, Driver distraction: a review of the literature, in *Distracted Driving*, ed. by I.J. Faulks, M. Regan, M. Stevenson, J. Brown, A. Porter, J.D. Irwin (Australasian College of Road Safety, Sydney, 2007), pp. 379–405
15. O. Dehzangi, C. Williams, Towards multi-modal wearable driver monitoring: impact of road condition on driver distraction, in *2015 IEEE 12th International Conference on Wearable and Implantable Body Sensor Networks (BSN)* (IEEE, New York, 2015), pp. 1–6
16. O. Omeni, A.C. Wai Wong, A.J. Burdett, C. Toumazou, Energy efficient medium access protocol for wireless medical body area sensor networks. IEEE Trans. Biomed. Circuits Syst. **2**(4), 251–259 (2008)
17. B.I. Kwak, J.Y. Woo, H.K. Kim, Driving dataset. PST 2016. http://ocslab.hksecurity.net/Datasets/driving-dataset
18. Mobile ID World, Keeping drivers safe with pupil biometrics (2016). https://findbiometrics.com/keeping-drivers-safe-withpupil-biometrics-301058/
19. Artificial intelligence helps to keep tired drivers awake (2017). http://www.digitaljournal.com/tech-and-science/technology/artificial-intelligence-helps-to-keep-tired-drivers-awake/article/499369
20. J. Lee, Omron integrating facial recognition into autonomous driving system (2016). http://www.biometricupdate.com/201610/omron-integrating-facial-recognition-into-autonomous-drivingsystem
21. Automotive. http://www.optalert.com/automotive
22. Vigo the science. https://www.wearvigo.com/science

23. E. Romera, V. Arroyo, L.M. Bergasa, Need data for driving behavior analysis? Presenting the public UAH-DriveSet, in *Proceedings of IEEE International Conference on Intelligent Transportation Systems (ITSC)* (IEEE, Rio de Janeiro, 2016), pp. 387–392
24. Sense holdings acquires exclusive sales, marketing and purchase rights for biometric patent to secure vehicles. http://www.theautochannel.com/news/2004/03/18/185331.html
25. K.A. Ishak, S.A. Samad, A. Hussain, A face detection and recognition system for intelligent vehicles. Inf. Technol. J. **5**(3), 507–515 (2006)
26. S. Ben-Yacoub, B. Fasel, *Fast Multi-Scale Face Detection* (IDIAP, Martigny, 1998)
27. P.N. Belhumeur, J.P. Hespanha, D.J. Kriegman, Eigenfaces vs. fisherfaces: recognition using class specific linear projection. IEEE Trans. Pattern Anal. Mach. Intell. **19**(7), 711–720 (1997)
28. Z. Zhu, F. Chen, Fingerprint recognition-based access controlling system for automobiles, in *2011 4th International Congress on Image and Signal Processing (CISP)*, vol. 4 (IEEE, New York, 2011), pp. 1899–1902
29. C.O. Folorunso, L.A. Akinyemi, A.A. Ajasa, K. Oladipupo, Design and development of fingerprint based car starting system, in Presentation made at 16th International Conference on Electronic Packaging Technology (ICEPT 2015), Exhibition on Power and Telecommunications. http://nieee.org.ng/portfolio/papers/view/Design_and_Development_of_Fingerprint_based_Car_Starting_System_(Folorunsho_C_et_al.pdf.Appendix: Springer-Author Discount

Internet of Things and Artificial Intelligence-Enabled Secure Autonomous Vehicles for Smart Cities

D. A. Janeera, S. Sheeba Rani Gnanamalar, K. C. Ramya, and A. G. Aneesh Kumar

Introduction

With the advent of several promising technologies such as Internet of Things (IoT) and artificial intelligence (AI), we have stepped into a new epoch of computing. There is a rapid growth in technology since the past few decades in both software and hardware paradigms [1]. With this technological progression, the hardware capacities are expanding continuously to meet the growing multidimensional requirements, while the device size is becoming consistently smaller every day.

IoT uses Internet- and sensor-enabled communication to bring together small objects commonly referred to as things. This facilitates wider and cheaper communication channels that can perform numerous tasks in any industry or smart city environment [2]. It enables collection and processing of information from sensors, actuators, storage devices, communication technologies, and other innovative services [3].

AI is a smart machine tool that offers huge prospects in the smart revolution. It enables collection of information, identification of suitable alternatives, selecting the best alternative, making smart choices, taking necessary actions, reviewing the choices, and smart prediction. The combination of IoT and AI brings together low-latency, resilient, and high-speed connectivity enabling completely smart autonomous automobiles in a smart city environment.

D. A. Janeera (✉)
Department of ECE, Sri Krishna College of Engineering and Technology, Coimbatore, India

S. S. R. Gnanamalar · K. C. Ramya
Department of EEE, Sri Krishna College of Engineering and Technology, Coimbatore, India

A. G. A. Kumar
Core Engineer-Switch Operations, Ericsson India Ltd, Coimbatore, India

M. Kathiresh, R. Neelaveni (eds.), *Automotive Embedded Systems*, EAI/Springer Innovations in Communication and Computing,
https://doi.org/10.1007/978-3-030-59897-6_11

This chapter focuses on examining IoT and AI and their latest contributions in the field of autonomous vehicles while not compromising the security of data in smart city environment. Research shows that over 90% of accidents are caused due to human errors [4, 5]. The use of autonomous vehicles offers a significant improvement in safety and reduces the rate of risk by 1000 times.

The key benefits of autonomous vehicles include low emission of dust, new market prospects, business opportunities and driver time release, reduced consumption of fuel, accident reduction, and increased safety of vehicles. Nevertheless, autonomous vehicles must use large amount of data retrieved from the devices and sensors. For applications related to entertainment and advanced driver assistance systems (ADAS), the complexity and rate of processing information is increasing (up to 1 gbps). Autonomous vehicles also reduce road accidents and death caused by accidents which sums up to over 3000 deaths per day.

To meet the growing software as well as hardware requirements, we use AI along with software, devices, actuators, and sensors to perform functions like a superhuman brain. The devices and sensors provide information such as virtual assistance, gesture and speech recognition, sentiment analysis, image recognition, driver monitoring, eye tracking, recommendation engines, voice search, cumulative mileage, deceleration, acceleration and vehicle speed, voice recognition, fuel consumption, navigation, motion detection, time, and date [6]. Annually, for every 1 lakh vehicles, around 100 TB of data is generated. Control, path planning, localization, sensor fusion, and computer vision are used in self-driving cars.

Artificial Intelligence

It is essential to distinguish between the concept of strong and weak artificial intelligence to understand the concept of artificial intelligence. The superhuman or human intelligence of a strong AI in all aspects remains a pure fiction till date. The most recent commercial applications use only the weak perception of AI. In this perception, AI can perform certain tasks such as dealing with complexity, probabilistic reasoning, understanding context, visual perception, and so on that require human capabilities in general [7].

The next generation will be greatly impacted by the application of AI technology in autonomous vehicles. With the diversified and ubiquitous implementations of AI, it is essential to comprehend the decision-based applications and their varied contexts. In order to make accurate decisions based on the context, it is essential to decode the concepts at a nonlinear relational level and face value [8]. Hence it is almost impossible to perceive a single model of AI as a solution for all applications. Since there are several probabilities for the decisions that can be taken by the machine based on the applications, it is difficult to fit it in a single framework.

Figure 1 provides the timeline of evolution of AI from the conceptual level to reality. Based on the applications, AI can be divided into machine learning, statistical

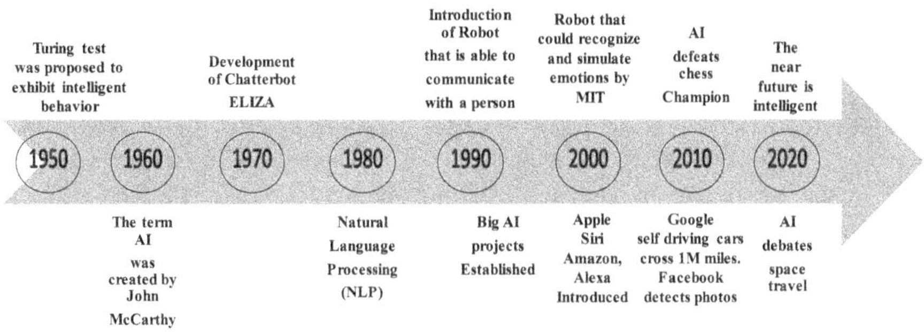

Fig. 1 Evolution of artificial intelligence (AI)

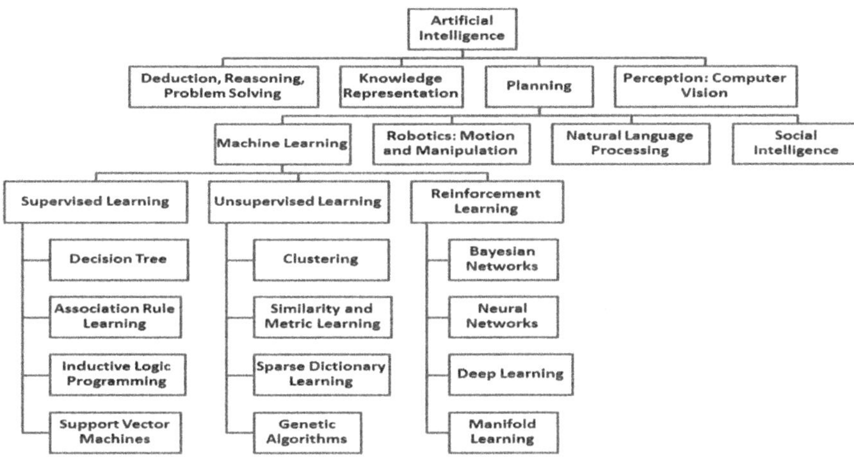

Fig. 2 Approaches of artificial intelligence

learning, and symbolic learning [natural language processing]. Figure 2 represents the various approaches of AI.

The mathematical analysis of data pertaining to problems and providing predictive functions is the key role of statistical learning. In order to build a model, a hypothesis is formed initially. The variables and their relationships are considered to create a rule-based programming [9]. Homoscedasticity, lack of multi-collinearity, normality, and such assumptions form the basis of statistical learning when small datasets with limited attributes are considered.

Numerical and analytical models can be created and automated by machine learning (ML) algorithms [10, 11]. In any specific task, these algorithms can improve the performance of the system. Without explicit planning of the search location or task to be accomplished, the data's hidden insights can be found by means of statistics, operations research, and heuristic methods. Machine learning is further categorized into reinforcement learning, supervised learning, and unsupervised learning.

Reinforcement learning helps in the advancement of AI in real-world machine learning with its efficient decision-making technique. Supervised learning uses the input and output information to create a model for prediction. The variables and the relationship between them are found by regression technique in order to predict the outcomes. Prediction of target classes for every data category is done by classification. Unsupervised learning uses only the input data to make assumptions. Cluster groups are created by classifying the data points using clustering algorithm [12].

Deep learning [13] is a flexible and powerful ML approach that uses the implication hierarchy for any representation. DL algorithms use an incremental approach for learning sophisticated features from specific information [14]. A domain expert is essential in several ML algorithms to identify the features applied, thereby reducing the data complexity and improving the visibility of patterns to learning algorithm.

Decision-making, perception, and conception form the base of deep learning. Numerous processing units and layers of neural networks are used to advance the training methodologies to assist in easy learning of complex patterns that contain enormous amount of data [15]. Object recognition, speech recognition, image, video labeling, and activity recognition are some of the complex applications that use DL. It is also used in transmission of inputs for NLP, perception of speech, and audio [16].

Internet of Things (IoT)

Initially, IoT and its applications were remarkable in the fields of manufacturing and business where applications were mostly dependent on machine-to-machine (M2M) communication. With the advancements in technology, IoT is finding its way into smart homes, offices, and smart cities, by transforming and connecting various technologies. Before the introduction of IoT, the early devices connected via the Internet include blogjets, pervasive computing, invisible computing, and ubiquitous computing [17].

Kevin Ashton in the year 1999 suggested the term Internet of Things (IoT). Over the last decade, the notion of the word "Things" has evolved. The concept of devices that can decode data without the intervention of humans remain constant. The interconnection of humans is made possible by means of the Internet at an exceptional pace and scale [18]. Creation of smart environments by means of interconnection of objects is the next wave of technology which has already started impacting the global population (Fig. 3).

The International Data Corporation (IDC), a technology analyst company, has predicted that by 2025, at least 41.6 billion devices will be connected through IoT. Currently, over 9 billion devices are connected, and the number exceeds the total world population count. For smart applications, it is possible to access huge datasets when large numbers of devices are connected [19]. In order to improve the control and efficiency, several industries such as automotive, manufacturing, mining, and agriculture have already incorporated IoT.

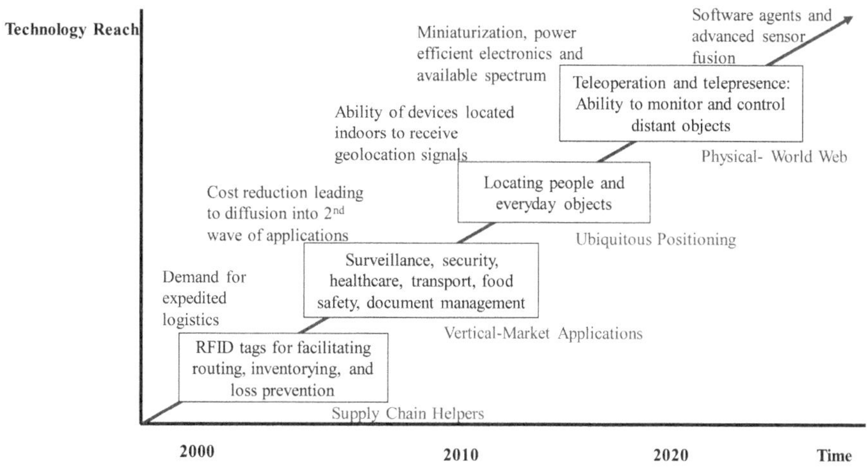

Fig. 3 Technology roadmap of Internet of Things

IoT has huge opportunities in the field of automotive and industrial equipment along with wearable and smart home devices. While using IoT for real-time applications such as smart traffic control, there will be strict requirements of delay when applied over communication networks with distributed sensing. These applications allow a tolerable delay for the system while providing an optimal solution in a permissive neighborhood [20]. In these applications, the performance is dependent on the delay of the optimization algorithm which has a built-in computational proficiency. The information sharing from the various sensors and uploading of information to the cloud is done over the IoT platform [21].

Vehicles, homes, and office environments are made more measurable and smarter with IoT. Getting information, playing music, setting time, and such operations are easier with smart devices like Google Home and Amazon's Echo [22]. Talking to visitors, seeing the visitors, and monitoring what is happening outside and inside the house even from a remote environment is made possible with home security systems. Smart light bulbs can change intensity based on the brightness of the surroundings, and smart thermostats can warm the house before you arrive home.

One of the major drawbacks of IoT is security. The sensors connected with IoT collects several data that are exceptionally sensitive. It is essential to ensure safety of this data in order to earn the trust of the consumer [23]. The security record of IoT is very poor. Data encryption and transmission is not given much importance in many devices. In the current scenario, ensuring the safety of this enormous amount of data is mandatory.

Autonomous Vehicles

In the 1930s, the concept of autonomous vehicles (AV) was introduced by writers of science fiction who visualized a new idea of the automotive industries—the self-driving cars. A vehicle that can operate on its own in a guided manner without human interaction is termed as an autonomous vehicle (AV). This mechanism uses computers for driving.

AV is besieged due to its various advantages when compared to conventional manual automobiles [24]. AVs expect an annual profit of around 7 trillion dollars by 2050. AV and autonomous driving are the most publicly followed and rigorously explored technology in the automotive domain. Due to the various complexities involved in the development stage, the complete adoption of AVs would take time and will be visible when we near the completion of development and overcome the existing drawbacks. This may take several simulations and iterations.

History of Autonomous Vehicles

Autonomous cars have swiftly turned from fantasy and science fiction to an on-road reality. Even though it seems like the technology evolved overnight, autonomous vehicles have emerged through a long path. This section consolidates the major milestones in the evolution of autonomous vehicles. Figure 4 represents the roadmap of a century of driverless cars from the 1920s to the 2020s.

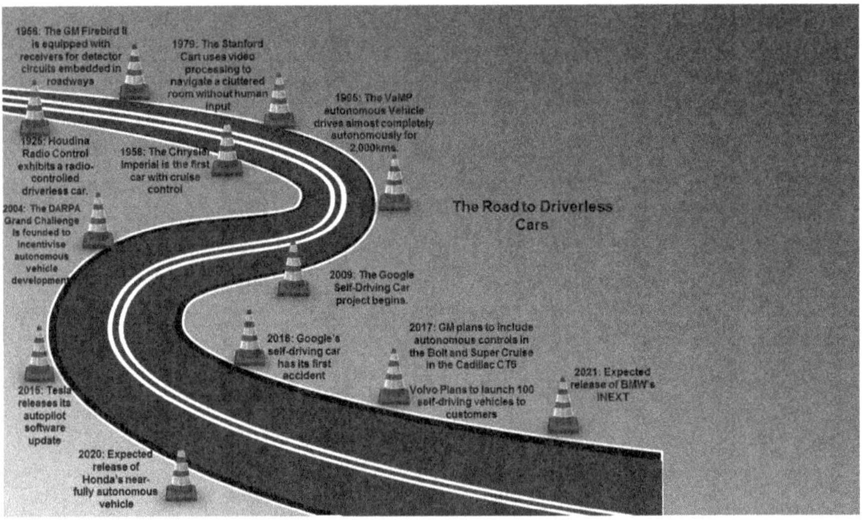

Fig. 4 Roadmap of autonomous vehicles

Francis Houndina, an inventor, demonstrated a driverless car which used radio control in the year 1925. John McCarthy in 1969 described AV as computer-controlled cars. Self-driving cars that used neural networks and real-time image processing technology were introduced during early 1990s by a researcher Dean Pomerleau from Carnegie Mellon [25]. A Grand Challenge was announced by DARPA in the year 2002 for building autonomous cars to travel through the Mojave Desert for a grand prize of 1 million dollars.

With several researchers working in the field of autonomous vehicles, self-parking systems using autonomous road technologies and other sensors were beginning to materialize. Several car manufacturers such as Toyota, Lexus, and BMW stated commercializing the self-parking feature. Waymo, a secret project for driverless cars, was started by Google in the year 2009. Google's autonomous cars have been tested for 300 thousand miles of accident-free driving in a short span. The prototype of this project was revealed by Google in the year 2014. Key automobile companies like BMW, Mercedes-Benz, Ford, General Motors, and so on started their own autonomous car projects by the year 2013. Nissan has planned on launching several autonomous automobiles in 2020. Other technology giants such as Uber, Tesla, and Apple are also exploring this technology.

An accident occurred when Tesla's autonomous car was driving in autopilot mode through Florida. The car hit a tractor trailer with 18 wheels and caused the death of the occupant. The car failed to apply the brakes on time when the trailer before it took a turn in front of it [26]. This created several debates about autonomous cars, the ethical and technical challenges.

Levels of Autonomous Vehicles

Driving automation is divided into six levels ranging from no automation to completely automated driverless vehicle. Figure 5 represents the various levels of autonomous vehicles and its range of automation.

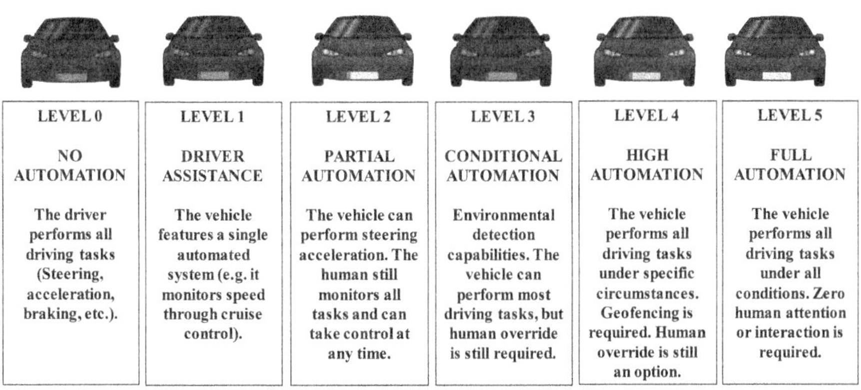

Fig. 5 Levels of autonomous vehicles

Level 0 automobiles are the present-day on-road vehicles which are completely manually controlled. Level 1 introduces the least automation level through the introduction of adaptive cruise control technology. Level 2 presents the advanced driver assistance systems or ADAS. Level 3 automobiles have a certain extent of decision-making capabilities based on sensing the surrounding environment. Level 4 vehicles can make interventions when it senses a human error or a system failure. Level 5 vehicles are completely independent and do not need any human attention [18].

Autonomous vehicles use machine learning, complex algorithms, actuators, sensors, and powerful processing units for software execution. A map of the surrounding environment is created and maintained by means of various sensors located in the vehicle.

Benefits and Challenges of Self-Driving Automobiles

In order to access path tracking, path planning, map building, and localization features, autonomous mobile navigation is required in autonomous vehicles. Detection of obstacles, classification, and avoidance is also to be enabled for safe transportation. The major issues of AVs include the following:

1. Lack of accuracy at the software level
2. Lack of precise and complete maps with innovative features and details
3. Unavailability of sensor-based prediction to identify the level of complexity of on-road situations

Due to the complexity of the performed tasks, the software, architecture, area of awareness, and sensors of AV have become complex. It is difficult for the sensors to perform quick processing of information and to discriminate between dangerous scenarios. In order to dodge obstacles and keep the vehicle on track, numerous devices and sensors are essential [25]. While driving, a large amount of data is generated by means of the sensors in order to create awareness of the vehicle and the situation of the surroundings in order to make suitable decisions.

It is necessary to have comprehensive software for an AV so as to combine the information of real-time response, failures, and situational awareness. The actions can be developed based on logic so as to ease the AV complexity. In addition to this, the volume of information at every state and the information retaining period can be minimized for further simplicity [23]. The performance of the AV is more deterministic when limited inputs are involved.

Automobiles have restricted maneuver and navigation capability. Hence reducing the amount of data is a challenging task. Due to this, several challenges are imposed to improve the efficiency of the AV system design by means of software and hardware architecture [27]. Information gathered from the various sensors are of different formats and have to be converted to a homogenous digital form to enable processing of data over a single platform.

Numerous challenges include acceptance and trust of consumers, lack of technological tools and industrial standards, regulatory factors, and so on. AVs have gone through several levels of refinement over the decades, and yet, at every level of progression towards autonomy, there are more complex challenges [15]. The technologies in prediction, control, planning, localization, and perception are to be improved. Some of the major factors that affect the operation of AVs are as follows: road, weather and traffic conditions, accident liability, radar interface, and big data analytics.

Roads can vary uncertainly from time to time. Tunnels with unclear signals, mountainous roads, potholes, undefined lanes, and so on affect the performance of AVs. Despite dreadful weather conditions, AVs are expected to operate without downtime or failure. AVs should also perform with the same efficiency even in unpredictable traffic conditions. In circumstances where AV causes accidents, it is not possible to hold anyone liable for the situation. The frequency range for communication between AVs will be limited even with the use of radar and lasers for navigation as it has to cater to the needs of several hundreds of on-road automobiles.

AVs should be capable of decision-making in real-time scenarios as well as making use of the training data and system [19]. With voluminous data used for this purpose, it has to be ensured that there is no compromise made on the speed of the system. Information procurement, storage, organization, and classification are the major considerations of big data in AV. Other smart approaches like Kernel-based learning, active learning, transfer learning, parallel and distributed learning, deep learning, and representation learning can be applied to improve the speed of processing the big data and to enhance the efficiency of the decision-making process.

Further, the AV can be enhanced by building an energy-efficient machine with alternative power sources, electric vehicles, and hybrids. The fuel rate of fuel consumption, emission of gasses, speed, acceleration, real-time location of the vehicle, and diagnostics of vehicular information should be uploaded in the network, enabling vehicular communication [27]. Environmental information such as parking information, eco-speed limits, eco-routes, safety messages, traffic signal messages, and real time traffic congestion are also to be communicated to the vehicle through proper means to ensure smooth operation of the AV [21].

Smart City

An urban area using various IoT sensors for data collection and efficient management of services, resources, and assets by means of the collected data is termed as a smart city. The information gathered from the assets, devices, and citizens are collected, processed, and examined to observe and manage community services, hospitals, libraries, schools, information systems, and crime detection.

In every area of working space and other environments, the use of information and communications technologies (ICTs) are transforming lives. There is a huge demand for digital and electronic communication technologies in cities and communities. Government systems are using embedded ICTs. The localization of ICTs and its relevant practices brings people and communities together and enhances the knowledge and innovation they provide [23].

A few prominent changes that have to be done to enable the use of autonomous vehicles in smart city infrastructure include lane marking, roadside sensors, and smart signage. Machine-readable markings on the road, sensors in lanes, curbs and sidewalks, signboards with transmittable embedded codes, and so on help in improving the infrastructure of the smart city.

Improvising the smart cities helps in developing future cities that may even use air taxis by means of autonomous drones that can carry a single passenger from one location to another [14]. A wide array of technologies, services, and applications can connect automobiles to its surroundings, infrastructure, other vehicles, services, applications, networks, and external devices. Applications assisting in global positioning system (GPS), telematics, remote diagnostics, roadside assistance, parking assistance, efficiency, and traffic safety are also included [28].

Data Security in Smart Environment

Exposure to the risks of cybersecurity and privacy of data are a major concern in autonomous automobiles. Key features such as information gathering, use of data, and decision-making are to be made secure in self-driving vehicles so as to achieve data privacy. Governments are enforcing several laws to overcome the issues of data theft in smart environment and automobiles. Table 1 represents the major security challenges in smart environment and their causes. By understanding the causes, it will be easy to overcome several of these issues.

The computerization of automobiles and generation of massive information is creating possibilities for unnecessary and unauthorized access to the information increasing the risks of cybercrimes [9]. The location of the vehicle, the commuters and their identities, time of travel, and other critical personal information can be hacked. Such cyberattacks can lead to disastrous situations and put the lives of the commuters of the car and its surrounding vehicles at risk.

Table 1 Major security challenges and causes

Sl. No.	Security challenge	Cause
1	Data loss	Unencrypted data
2	Eavesdropping	Insecure communication channel between the sender and receiver
3	Compromise on confidentiality and integrity	User ID, password, and such data authentication credentials are missing
4	Compromise on availability	Hardware and sensor issues

IoT- and AI-Enabled Autonomous Vehicles

This system uses several AV objective sensors like ultrasonic sensor, cameras, radar, light detection and ranging [LiDAR] sensor, and dedicated short-range communications (DSRC) sensor. Other sensors include AV pose sensors, GPS, wheel odometry, accelerometers, and gyroscopes. Fusing the information that is sensed locally in such a way that it supports the decision-making of the vehicle is a major challenge in understanding and reconstruction of the AV environment [29].

Several software like sensor fusion are developed for the purpose of combining the information from numerous sensors for decoding and enhancing the performance of the system. The data from the independent sensors can be combined, and the orientation and position of the vehicle can be calculated by proper data fusion. The techniques commonly used for the fusion of data are estimation using Kalman filter or weighted average; classification using decision tree, support vector machines, discriminant analysis, k-nearest neighbor, and density estimation; inference using Dempster-Shafer evidential reasoning and naive Bayesian inference; and artificial intelligence by means of artificial neural networks and fuzzy logic [5].

Figure 6 represents a typical secure smart car integrated with an array of sensors, AI, and IoT platforms that can simultaneously support millions of devices and generate voluminous data that can be encrypted, transmitted, and processed in a cloud computing environment. A typical AI-IoT-AV consists of four main components. The primary and fundamental component is the hardware and sensor devices that collect information from the surroundings in different formats. Cellular technology,

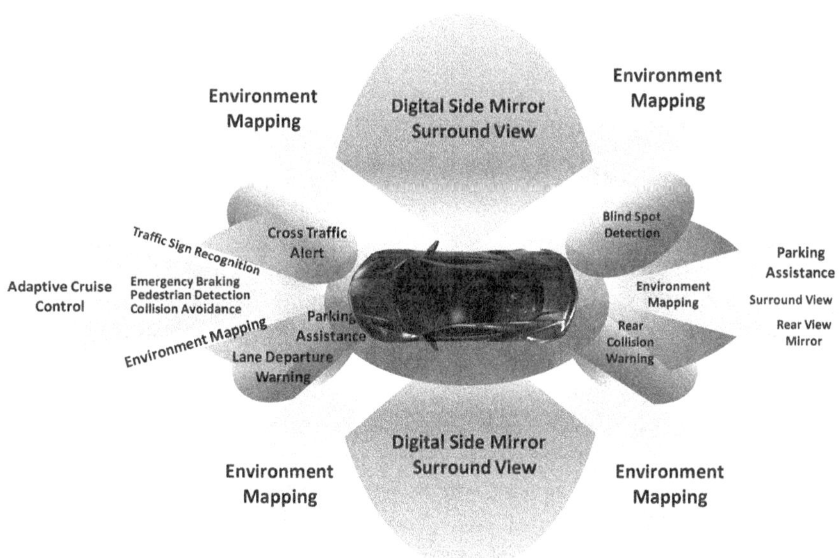

Fig. 6 Secure smart car

wireless technology, or Wi-Fi based communication networks form the secondary component of the system that helps in the transfer of data from the sensors. The large volume of data generated is to be transmitted and stored in an environment that can keep up with the velocity and volume. For this purpose, big data serves as the tertiary component. The fourth and foremost component where the data processing is done makes use of cloud computing platforms for storage, analysis, and processing of the data. The digital surround view camera offers a 360° view of the AV that leaves no blind spots.

Conventionally, the cloud environment hosts the IoT applications and delivers the decisions and feedbacks to the physical systems. The monitoring tools as well as the software components will be hosted in the cloud in a typical autonomous vehicle providing a central management system [24]. The AV interacts with the roadside device, sensing device, network infrastructure, personal device, commuter, and the vehicle. Exchange of information such as alerts, road conditions, and other parameters of the physical system can be done between the nearby vehicles as well. Commuters inclusive of drivers, passengers, and their personal devices can be connected to the AV system for data transfer. The entire communication depends on the optimal communication network that allows secure and fast transfer and exchange of data in the smart environment.

Information regarding the environmental variables, health levels of the commuters, parameters of the vehicles, and so on is collected from the actuators and sensors. The weather condition, noise level, pollution, heart rate of the commuters, blood pressure, vehicle temperature, fuel consumption, tire pressure, and several other important data can be sensed by the sensing devices. The radars, information screens, and traffic lights can disseminate road- and traffic-related information and intimate about possible detours and accident occurrences.

RSA Encryption Algorithm

In order to ensure the safety of information, several algorithms were developed enabling data encryption. The Elliptical Curve Cryptography (ECC) algorithm, Advanced Encryption Standard (AES) algorithm, Data Encryption Standard (DES) algorithm, and Ron Rivest, Adi Shamir and Leonard Adleman (RSA) algorithm are compared. The RSA algorithm is found to be ideal among the compared algorithms. It is further enhanced for applications the require encryption of the information transferred by Artificial Vehicles [30] (Table 2).

The Big O notation has been used for expressing the complexity of the algorithm for analysis purpose. The algorithm runtime is classified into its principal parts, and lower-degree polynomials and coefficients are dropped while using this notation. An algorithm executing a loop which has the length n-3 times is described by $O(n)$ rather than $3n$. For every couple of years, the key lengths of RSA are increased to ensure that there is no compromise on the improved factoring algorithms. For generating key pairs greater than the length of 1024 bits, the performance of the RSA

Table 2 Comparative chart of encryption algorithms

Algorithm	ECC	AES	DES	RSA
Developer	Neal Koblitz Victor S. Miller	Rijman, Joan	IBM-75	Amazon
Execution time	Fast	Faster	Slow	Fastest
Security rate	Good	Good	Insufficient	Excellent
Block size	Variant	128 bits	64 bits	Variant
Key length	135 bits	256, 192, and 128	56 bits	$N = p \times q$

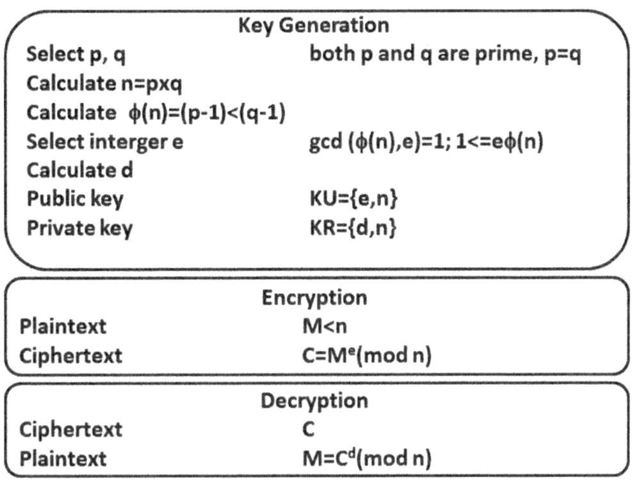

Fig. 7 Generation of RSA key

algorithm is to be investigated. The overall implementation efficiency of the system can be improved with certain special features of the crypto coprocessors.

Key creation, dissemination, encryption, and decryption are the four steps involved in RSA algorithm. It comprises a public and a private key. Encryption of messages is done by means of the public key and is recognized by everybody. Decoding the message encrypted by the public key is done with the use of the private key for a limited time duration. Figure 7 represents the key generation using RSA algorithm. p and q are nonidentical prime numbers that are kept secret. Key length is given by n which is the product of p and q [10].

Edge Computing Model

It is necessary to use offloading data processing in the present-day IoT platforms for autonomous vehicles for real-time processing of data with low latency. The cloud facilitates the development and perpetuation of applications and its relevant information while providing access to storage and computational assets. Edge computing allows the computation of information in a close setting which may be prefabricated, minor data centers. This helps in making quick decisions and offering quick responses to the AV system.

The utilization of resources is optimized by cloud computing environment; there are several challenges in accommodating applications for autonomous vehicles. The major challenges include difficulty in integration of data along with the software and hardware components, increased workload due to the voluminous data, physical location of the cloud, reduced throughput and latency, and increased cost and energy consumption [24].

Figure 8 represents an IoT-based AV ecosystem with edge computing technology that has a centralized cloud computing environment and distributed edge computing networks that access information from smart homes, smart cities, and autonomous vehicles. Various hardware such as GPU, FPGA, multicore processors, computing boards like ARM, Raspberry Pi, BeagleBoard, and so on can be used as edge computing devices.

For applications that require fixed and configurable hardware resources, edge computing devices are used. Here, the resources allocated are vague and cannot be controlled by the user. It is possible to integrate the mobile IoT nodes for AVs using edge computing. In close proximity, multiple edge devices can build subsystems for exchange of information. The pay-as-you-go model is facilitated by cloud for the edge users. Energy efficiency of edge devices are critical as they can be powered by a battery, while the cloud is provided with a constant power source [26].

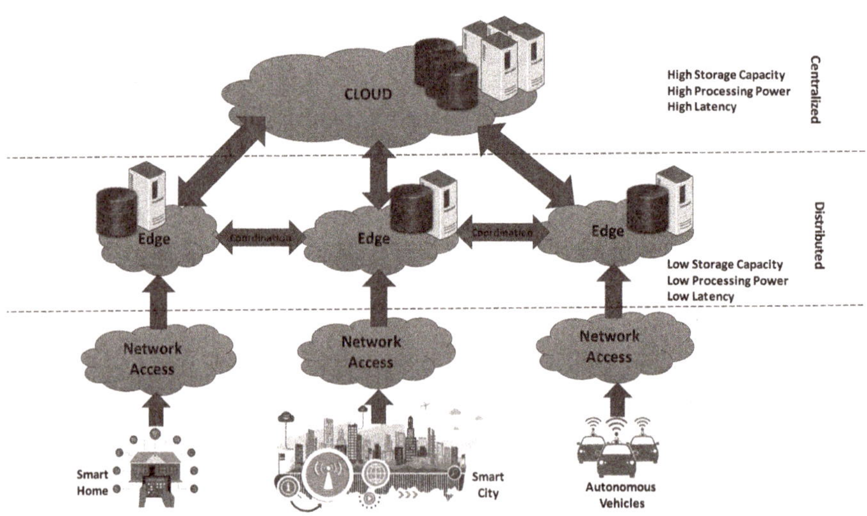

Fig. 8 IoT-based AV ecosystem with edge computing technology

Results and Discussion

The system's privacy architecture decides the successful outcomes of the operation of the autonomous vehicles. In AVs, the success of the system improves with reduction in the risk factor that leads to more frequent and confident usage of the system. There will be an increase in the privacy and security of information. This increases the system performance proportionally and reduces the possibilities of accident occurrences.

Cybersecurity is given a high priority in this system, and hence the system is secure from cyberattacks. The manufacturers and commuters of such AVs will not be affected by the data security and privacy issues. Furthermore, safety is improved by using location tracking feature, and the gathered information can prove beneficial in critical circumstances.

The major acceptance postulate provides over 90% success against cyberattacks. Tests have been conducted to find the accuracy of the postulate to produce four types of use case scenarios, namely, true positive, false positive, true negative, and false negative. The accuracy rate is dependent on appropriate encryption, data storage, and level of data security from cyber attacks.

When considering a genuine esteem, the maximum closeness of the measured actual estimation depicts the accuracy of the system according to ISO 5725-1. A precise error segment and an irregular blunder part are included at a point where the term is linked to similar estimation measure sets. In the arrangement of results, the closeness of agreement is represented by accuracy, and in the estimation of results, the closeness of the mean is termed as genuineness (Fig. 9).

A test capability to distinguish between the encrypted and stored information correctly is termed as the accuracy of the test. The true positive, true negative, and proportion of the cases are evaluated to estimate the accuracy of the test. The formula for accuracy can be represented as

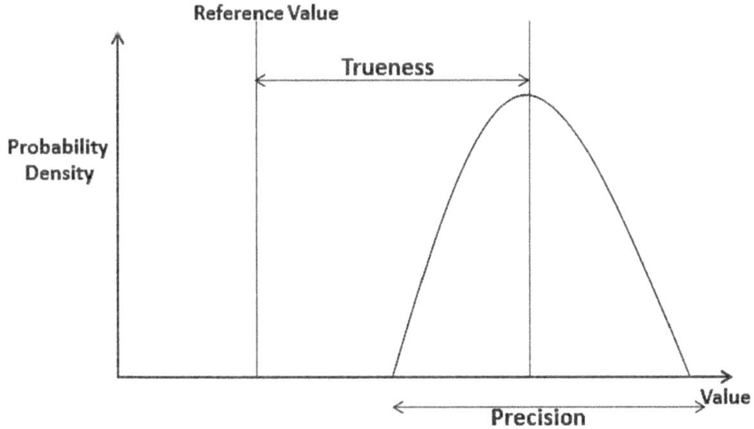

Fig. 9 Accuracy factors

$$\text{Accuracy} = \frac{\text{True Positive} + \text{True Negative}}{\text{True Positive} + \text{True Negative} + \text{False Positive} + \text{False Negative}}$$

The conventional model of cloud used for data analytics and storage is modified to accommodate edge computing to improvise the AI model of autonomous vehicles. The database system and the control mechanisms are implemented in the cloud in traditional models. This model is improved by dividing the planning and process sections into independent modules that can be handled collaboratively by the cloud and the edge.

Decision-making and pre-processing are done in the edge node to which data is provided by edge computing system of the artificial intelligence-based autonomous vehicles. The decision-making process that are less time-sensitive are sent for offline processing to the cloud as IoT sensor data using the edge nodes, while the rest of the information are analyzed using the edge locally. Crash avoidance, detection of obstacles, and such decisions that are time-sensitive are performed within a short span by the edge node.

In order to provide a better driving experience and improved road safety measures, the information regarding the diving pattern, traffic, and road are analyzed by the cloud. Based on the requirements of the customers, regulations, rules, and policies, updating the edge node can be done for implementation of the AI models. The voluminous data acquired from AV IoT sensors undergo pre-processing, filtering, and noise removal processes at the edge node before being offloaded into the cloud environment. These processes reduce the data size to a certain extent, thereby saving considerable amount of cost and bandwidth.

Conclusion and Future Scope

With the advancements in autonomous vehicles, adoption of smart technologies and techniques for improving the quality and performance of decision-making is made possible. Creation of robust and dynamic control systems is made possible in AVs with embedded systems that offer high performance through the integration of IoT and AI. Security, bandwidth, and latency issues of the existing smart vehicles can be overcome using the proposed RSA encryption algorithm and edge computing model at the software level.

Autonomous vehicles can reduce global emission of carbon dioxide by 80%. It can help in easy identification and allocation of parking spaces, improve living standards, reduce the cost of transportation by 40%, and decrease congestion due to traffic to a great extent. AVs should be agile and transformative to probable changes. Modularity, digital traces, smart systems, reprogrammable capacity, connectivity, decoupling, and homogenization factors are to be equipped.

RSA algorithm offers excellent data encryption and decryption features enhancing the privacy and safety of the data. This technique is also most suitable for cloud computing systems with multiple cloud networks and has proved to be efficient in banking and online shopping applications. This framework also prevents digital assaults as the protection and information security levels are greater. Future work can be performed towards enhancing the smart vehicles with these security models as the core system. Under chaotic situations and environment, the functions of AI can be further refined.

References

1. K.-F. Wu, L. Sasidharan, C.P. Thorc, S.-Y. Chena, Crash sequence based risk matrix for motorcycle crashes. Accident Anal. Prevent. **117**, 21–31 (2018)
2. R. Lanctot, Accelerating the future: The economic impact of the emerging passenger economy (2017). Retrieved from https://newsroom.intel.com/newsroom/wpcontent/uploads/sites/11/2017/05/passenger-economy.pdf
3. S.A. Beiker, Legal aspects of autonomous driving. Santa Clara L Rev. **52**(4), 1145–1156 (2012)
4. S.J. Russell, P. Norvig, *Artificial Intelligence: A Modern Approach*, 3rd edn. (Prentice Hall, Englewood Cliffs, 2010)
5. B.V. Philip, T. Alpcan, J. Jin, M. Palaniswami, Distributed real-time IoT for autonomous vehicles. IEEE Trans. Ind. Inf. **15**(2), 1131–1140 (2018)
6. A. Kankanhalli, Y. Charalabidis, S. Mellouli, IoT and AI for smart government: a research agenda. Gov. Inf. Q., 304–309 (2019)
7. S. Soomro, M.H. Miraz, A. Prasanth, M. Abdullah, Artificial intelligence enabled IoT: traffic congestion reduction in smart cities. IET Smart Cities Symp., 13–16 (2018)
8. H. Lu, Q. Liu, D. Tian, Y. Li, H. Kim, S. Serikawa, The cognitive internet of vehicles for autonomous driving. IEEE Network **33**(3), 65–73 (2019)
9. J. Pan, Z. Yang, Cybersecurity "challenges and opportunities in the new" Edge Computing+ IoT "World", in *Proceedings of the 2018 ACM International Workshop on Security in Software Defined Networks & Network Function Virtualization* (2018), pp. 29–32
10. F. Al-Turjman, H. Zahmatkesh, R. Shahroze, An overview of security and privacy in smart cities'. IoT Commun. **2019**, e3677 (2019)
11. L. Trotter, M. Harding, M. Mikusz, N. Davies, IoT-enabled highway maintenance: understanding emerging cybersecurity threats. IEEE Pervas. Comput. **17**(3), 23–34 (2018)
12. J. Joy, M. Gerla, Internet of vehicles and autonomous connected car-privacy and security issues, in *2017 26th International Conference on Computer Communication and Networks (ICCCN)* (IEEE, 2017), pp. 1–9
13. J. Cui, L.S. Liew, G. Sabaliauskaite, F. Zhou, A review on safety failures, security attacks, and available countermeasures for autonomous vehicles. Ad Hoc Netw. **90**, 101823 (2019)
14. D. Lu, D. Huang, A. Walenstein, D. Medhi, A secure microservice framework for iot, in *2017 IEEE Symposium on Service-Oriented System Engineering (SOSE)* (IEEE, 2017), pp. 9–18
15. L.N. Long, S.D. Hanford, O. Janrathitikarn, G.L. Sinsley, J.A. Miller, A review of intelligent systems software for autonomous vehicles, in *2007 IEEE Symposium on Computational Intelligence in Security and Defense Applications* (IEEE, Piscataway, 2007), pp. 69–76
16. M. Hengstler, E. Enkel, S. Duelli, Applied artificial intelligence and trust—the case of autonomous vehicles and medical assistance devices. Technol. Forecast. Soc. Change **105**, 105–120 (2016)
17. S. Poudel, Internet of Things: underlying technologies, interoperability, and threats to privacy and security. Berkeley Technol. Law J. **31**(2), 997–1022 (2016)

18. A. Alheeti, M. Khattab, K. McDonald-Maier, Intelligent intrusion detection in external communication systems for autonomous vehicles. Syst. Sci. Contr. Eng. **6**(1), 48–56 (2018)
19. M. Wolf, A. Weimerskirch, C. Paar, Secure in-vehicle communication, in *Embedded Security in Cars* (Springer, Berlin, 2006), pp. 95–109
20. S. Parkinson, P. Ward, K. Wilson, J. Miller, Cyber threats facing autonomous and connected vehicles: future challenges. IEEE Trans. Intell. Transport. Syst. **18**(11), 2898–2915 (2017)
21. S. Tayeb, M. Pirouz, G. Esguerra, K. Ghobadi, J. Huang, R. Hill, D. Lawson, et al., Securing the positioning signals of autonomous vehicles, in *2017 IEEE International Conference on Big Data (Big Data)* (IEEE, Piscataway, 2017), pp. 4522–4528
22. D. Grewe, M. Wagner, M. Arumaithurai, I. Psaras, D. Kutscher, Information-centric mobile edge computing for connected vehicle environments: Challenges and research directions, in *Proceedings of the Workshop on Mobile Edge Communications* (2017), pp. 7–12
23. S. Garg, A. Singh, S. Batra, N. Kumar, L.T. Yang, UAV-empowered edge computing environment for cyber-threat detection in smart vehicles. IEEE Netw. **32**(3), 42–51 (2018)
24. M. Bojarski, D. Del Testa, D. Dworakowski, B. Firner, B. Flepp, P. Goyal, L.D. Jackel, et al., End to end learning for self-driving cars. arXiv preprint arXiv:1604.07316 (2016)
25. A. Hars, Self-driving cars: the digital transformation of mobility, in *Marktplätze im Umbruch* (Springer, Berlin, 2015), pp. 539–549
26. A. Ndikumana, N.H. Tran, C.S. Hong, Deep learning based caching for self-driving car in multi-access edge computing. arXiv preprint arXiv:1810.01548 (2018)
27. D. Phan, A. Bab-Hadiashar, C.Y. Lai, B. Crawford, R. Hoseinnezhad, R.N. Jazar, H. Khayyam, Intelligent energy management system for conventional autonomous vehicles. Energy **191**, 116476 (2020)
28. P.A. Hancock, Some pitfalls in the promises of automated and autonomous vehicles. Ergonomics **62**(4), 479–495 (2019)
29. S. Gong, A. Zhou, S. Peeta, Cooperative adaptive cruise control for a platoon of connected and autonomous vehicles considering dynamic information flow topology. Transport. Res. Record **2673**(10), 185–198 (2019)
30. K. Mayilsamy, N. Ramachandran, V.S. Raj, An integrated approach for data security in vehicle diagnostics over internet protocol and software update over the air. Comput. Electr. Eng. **71**, 578–593 (2018)

Index

M. Kathiresh, R. Neelaveni (eds.), *Automotive Embedded Systems*,
EAI/Springer Innovations in Communication and Computing,
https://doi.org/10.1007/978-3-030-59897-6

D

V

W

Y